35th Eg-MRS International Conference

IOP Conference Series: Materials Science and Engineering
Volume 1171

Cairo, Egypt
3-4 July 2021

Editors:

Arafa Hussien Aly

ISBN: 978-1-7138-4631-4
ISSN: 1757-8981

Printed from e-media with permission by:

Curran Associates, Inc.
57 Morehouse Lane
Red Hook, NY 12571

Some format issues inherent in the e-media version may also appear in this print version.

Institute of Physics
Dirac House, Temple Back
Bristol BS1 6BE UK

Phone: 44 1 17 929 7481
Fax: 44 1 17 920 0979

techtracking@iop.org

Additional copies of this publication are available from:

Curran Associates, Inc.
57 Morehouse Lane
Red Hook, NY 12571 USA
Phone: 845-758-0400
Fax: 845-758-2633
Email: curran@proceedings.com
Web: www.proceedings.com

TABLE OF CONTENTS

IOP Conf. Series: Materials Science and Engineering 1171 (2021) 011001 doi:10.1088/1757-899X/1171/1/011001

Proceeding of 35th International Conference of Material Sciences and Its Applications (35th Eg-MRS) 2021

Preface

Arafa H.Aly
Physics department, Faculty of Sciences, Beni Suef University

It is with great pleasure that I introduce the proceedings of the 35th Eg-MRS International Conference, 3-4 July 2021, Cairo, Egypt. This conference was dedicated to technical issue related to material sciences research and its applications.

The objectives of the conference were not only to provide opportunities to the international scientists to present state of the art research in material sciences but also to provide a forum to discuss ideas for future directions in the field and explore possibilities for collaborative research projects.

The conference program included keynote, oral, and poster presentations from scholars working in the areas of materials science and engineering from all over the world. It covered recent trends and progress made in the field of material sciences. Professors from Egypt, UK, USA, Canada, Netherlands and France were invited to deliver keynote lectures regarding the latest information in their respective areas of expertise.

All papers were subjected to peer-review by conference committee members and international reviewers. The papers were selected based on their quality and their relevance to the conference. The proceedings present recent advances in the fields of materials science and engineering and manufacturing as well as in other related areas including materials properties, measuring methods, analytical and numerical modeling and simulations among others. I would like to express special gratitude to members of the conference committee and organizers of the conference. I would also like to thank the reviewers for their valuable time and advice which helped in improving the quality of the papers selected for presentation at the conference and for publication in the proceedings. Finally I want to thank the authors, the members of the organizing committee, the reviewers, the chairpersons, sponsors, and all other conference participants for their support of Eg-MRS 2021.

Conference Chairman

Prof. Dr Arafa Hussien Aly
Beni-Suef University, Egypt
Eg-MRS Chiarman

Coordinator (Eg-MRS)

Prof Dr.Hassan Talaat (Ain Shams University, Egypt)
Prof. Dr. Hassan Afify (*National Research Center, Egypt)*
Prof. Dr.Talaat Meaz (*Tanta University, Egypt*)
Prof Dr. Amer El-Korashy (Eg-MRS)
Prof Dr. Amal Kasry (*The British University in Egypt)*

Organizing and Scientific Committee

Prof Dr. Ahmed Ramadan	*(Helwan University*
Prof Dr. Sohair Negm	*(Benha University)*
Prof Dr. Hamed Hassan	*(Alexandria University)*
Dr. Hassan Nageh	(BUE)
Prof Dr. Samia A Saafan	*(Tanta University)*
Prof Dr. Gamal Turky	*(NCR Cairo)*
Prof Dr. Mohamed Shaban	(Beni-Suef Univ.)
Prof Dr. Ahmed Ghitas	(NRIAG)
Prof Dr. Soltan Eid A. Soltan	*(Helwan Uiversity)*
Dr. Mohmed Saleh	(MUST)
Mr. Zaky A Zaky	(Beni Suef University)

Editors and Reviewers

Editor Prof Dr.Arafa Hussien Aly, Beni-Suef Univerrsity, Egypt

Reviewers

1. Prof Dr.Hassan Talaat, Ain Shams University, Egypt
2. Prof Dr. Hassan Afify, National Research Center, Egypt
3. Prof Dr.Adel Phillips, Ain Shams University, Egypt
4. Prof Dr.Hamdy Farghaly, El-Minia University, Egypt
5. Prof Dr.Sedky Sodky Abolila , Sohag University, Egypt
6. Prof Dr. Ahmed Ghitas, NRIAG, Helwan, Egypt
7. Prof Arafa H. Aly, Beni Suef University, egypt
8. Prof. Tarek El-brolossy, Ain Shams University, Egypt
9. Dr.Numan Arshid, Malya University, Malysia
10. Prof Mostafa F.Eissa , Beni Suef University, Egypt
11. Prof Mahi Singh, Canada
12. Prof Saeed AbdelGhany, Benha University, Egypt
13. Prof Sahar Elnagar, Cairo University, Egypt
14. Dr.Hassan Nageh, BUE, Egypt

IOP Conf. Series: Materials Science and Engineering 1171 (2021) 011002 doi:10.1088/1757-899X/1171/1/011002

Peer review declaration

All papers published in this volume of *IOP Conference Series: Materials Science and Engineering* have been peer reviewed through processes administered by the Editors. Reviews were conducted by expert referees to the professional and scientific standards expected of a proceedings journal published by IOP Publishing.

35th International Conference, 35th Eg-MRS 2021

Arafa H.Aly

Physics Department, Faculty of Sciences, Beni-Suef University, Egypt

arafaaly@aucegypt.edu

Abstract. The overview describes the main directions and results of the 35th Eg-MRS International Conference, 35th Eg-MRS-2021 held in Cairo, Egypt, 3-4 July 2021. It gives the details about the participants and the proceedings.

The Volume contains Proceedings of the 35th Eg-MRS International Conference, Eg-MRS-2021:

Modernization, Innovations and Progress" which was held in Cairo, Egypt in July 3-4, 2021. The Conference held online using Zoom due to COVID-19.

The purpose of the Conference is to share the results and prospects of the achievements in using advanced scientific, innovative and information technologies in materials sciences and its applications.

In fact, this is the third year of publishing a special issue of Eg-MRS international conferences under the umbrella of IOP material sciences and engineering. The first one was in November 2019; Hurghada, Egypt [1]. The second one was in August 2020 which was online due to COVID-19 [2].

- Type of peer review: Double-blind
- Conference submission management system: Online: https://cutt.us/mo8G5
- Number of submissions received: 45
- Number of submissions sent for review: 25
- Number of submissions accepted: 14
- Acceptance Rate (Number of Submissions Accepted / Number of Submissions Received X 100):
- Average number of reviews per paper: 56%
- Total number of reviewers involved: 14
- Any additional info on review process: No

- Contact person for queries: Prof Arafa H.Aly; arafaaly@aucegypt.edu

All accepted papers were at a reasonable level of language, and it was requested from the authors more than once to improve the language to make it readable by the arbitrators. Many reviewers asked the authors to enhance the language more than once. The authors of the accepted papers did the possible to polish the language and the required requests of them to enhance the language within the available limits.

[1] Arafa H. Aly, 2020 *IOP Conf. Ser.: Mater. Sci. Eng.* **762, 0122005**

[2] Arafa H. Aly, 2020 *IOP Conf. Ser.: Mater. Sci. Eng.* **956 011002**

IOP Conf. Series: Materials Science and Engineering 1171 (2021) 012001 doi:10.1088/1757-899X/1171/1/012001

Spectroscopic study of Cu, Mn, Cd as heavy metals in agricultural samples

Noura Fayek[1], Walid Tawfik[2,*], Ahmed Khalafallah[3], Sawsan Hamed[1] and Wafaa Mousa[1]

[1]Department of Physics, Faculty of Women for Arts, Science and Education, Ain Shams University

[2]National Institute of Laser Enhanced Sciences (NILES), Cairo University, Cairo, Egypt

[3]Department of Botany, Faculty of Women for Arts, Science and Education, Ain Shams University

*corresponding author E-mail: walid_tawfik@niles.edu.eg

Abstract: Heavy elements represent a source of toxicity when accumulate in the soil and be transmitted to plants, animals and humans through food chains that may affect the human and the animal health. In this study, environmental samples, vegetable samples (Cabbage, Mallow, Turnip) and soil, were collected from Qaha in Qaliubia in Egypt then subjected to investigate the availability of toxic metals such as copper, manganese and cadmium using flame atomic absorption spectroscopy (FAAS) technique. We have found Cu, Mn and Cd in cabbage with concentrations around 25.28 ±1.263 ppm, 103.83 ±5.19 ppm and 0.792 ±0.0396 ppm, respectively. The Cu, Mn and Cd concentrations in Mallow were found to be 35.26 ±1.76 ppm, 142.72 ±7.14 ppm and 1.3 ±0.0649 ppm respectively. Also, the concentrations of Cu, Mn and Cd in Turnip have achieved 29.29 ±1.45 ppm, 79.33 ±3.97 ppm and 1.99 ±0.099 ppm respectively. For soil, the concentrations of Cu, Mn and Cd were 2.4 ±0.12 ppm, 4.8 ±0.24 ppm and 1.75 ±0.088 ppm respectively. It was concluded that the pollution index values for Cu, Mn and Cd in soil were 0.024 ±1.2 $\times10^{-3}$, 0.01 ±5.49 $\times10^{-4}$ and 0.58 ±0.029 respectively. The pollution index values in cabbage for Cu, Mn and Cd were 0.346 ±0.017, 15.708 ±0.785 and 3.96 ±0.198 respectively. The pollution index values for Cu, Mn and Cd in Mallow were 0.073 ±0.024, 21.59 ±1.08 and 6.5 ±0.32 respectively. The pollution index values for Cu, Mn and Cd in Turnip were 0.401 ±0.019, 12 ±0.6 and 9.95 ±0.5 respectively. The observed new results are important to monitor the environmental pollution in the studied samples which have an impact on human health.

Keywords: Environmental samples, heavy elements, toxic metals, FAAS

1. Introduction

Food has been an integral part of human diet since the dawn of time; it is a rich source of enzymes, vitamins, and minerals. Modern diet, on the other hand, emphasises the importance of vegetables and fruits in improving life satisfaction [1]. Vegetables are the most polluted food [2], absorbing heavy metals and accumulating them in edible and non-edible pieces in concentrations large enough to cause health problems in humans [3, 4]. Heavy metals in high amounts harmed the immune, cardiovascular, renal, and neurological systems, as well as causing bone disorders and other health issues [5]. Heavy metals in contaminated soils have been linked to an increased risk of cancer [6, 7]. Metal pollution in some Egyptian landscapes is still widely common, resulting in environmental toxicity [1, 8].

Heavy metal pollution in soils, such as Cd, Ni, Zn, Pb, and Cu, has risen sharply in recent decades [9] as a result of mining, smelting, refining, agricultural fertilisers and pesticides, urban waste, traffic pollutants, and industrial effluents. Heavy metal pollution in soils is now common [11]. Land pollution caused by heavy metals has a significant negative impact on the climate and ecosystem all over the world [12, 13]. Food waste is caused by heavy metals dispersing in irrigated soils and growing plants, which can be toxic to humans and animals [2]. Heavy metals in effluents are poorly absorbed in water and thus cannot be degraded; as a result, they appear to settle in soils and therefore in plants [14–18]. Heavy metals also persist in the soil, where they leach into the groundwater, where they can stimulate antioxidant enzymatic activities in plants or become adsorbed by solid soil particles [19]. Because of the continuing rise in metal quantities in the atmosphere, most developed countries' governments have changed their laws to reduce the prevalence of these elements in the environment. For instance, it is well established that lowering lead levels in gasoline has resulted in a significant reduction of this metal in the atmosphere [20–23]. Several studies have shown that children's exposure to soils can result in their ingestion of a significant number of toxic elements [24]. Furthermore, children are at higher risk than adults due to their decreased susceptibility, higher ingestion rates from the gastrointestinal tract, and unique practices (hand–mouth operation, outdoor sports, poor hygienic habits, etc.) [24, 25].

There are advanced techniques that have been used for the analysis of soils: impedance spectroscopy [26] which has been used for soil analysis [27], high sensitivity, low detection limits, wide dynamic range, and high-speed multi-element analysis, inductively coupled plasma optical emission spectrometry (ICP-OES) [28] and laser spectroscopic techniques [29–31] (like laser- induced breakdown spectroscopy (LIBS) [32–38].

Atomic Absorption Spectrometry (AAS) is a method for calculating the ingestion of compounds introduced in soil experiments by measuring the amount of radiation absorbed by the compound portion of intrigue. This is followed by an examination of the spectra produced when the sample is irradiated. It is based on the Beer-Lambert law, which uses an atomic absorption method to calculate vitality as photons of light absorbed by the sample. A detector matches the wavelengths of light emitted through the sample to the wavelengths that went through the sample initially. The change of wavelength absorbed is then integrated by a signal processor, resulting in peaks in energy absorption at discrete wavelengths in the readout [39].

This work is aimed to determine the concentrations of heavy and toxic metals such as copper, manganese and cadmium in environmental samples: vegetable samples (Cabbage, Mallow, Turnip) and soil precisely.

2. Materials and methods

2.1 *The area of study*

The studied area is located at Qaha, Al- Qaliubiya Governorate, Egypt. Qaha city located 25 km north of Cairo. It exists in the Delta's rich agriculture in the south, which is well-irrigated by canals that branch off the Delta Barrage. Qaha city is a well-known industrial city with a quit number of factories and companies such as chemical industries, batteries, war products, shoe varnishes, aluminum tubing and fireworks.

2.1.1 *Sample collection and sample preparation*

Vegetable samples (cabbage, mallow, turnip) and soil were collected in winter 2019. The soil sample was collected using a shovel and an axe at a depth of 0-30 cm. The vegetable sample were divided into roots, stems and leaves and then washed by bi-distilled water. The samples were left to dry by air and then dried by an oven at 60°C until the weights remain constant, ground using a mortar and pestle. After that, they were stored in vacuum bags.

The digestion method: 0.2 gm of each sample were digested in 10 ml acid mixture of HNO_3: $HClO_4$ (5:1 v/v) for 8 hours at 80°C, then the digests were filtered and diluted to 50 ml with double de-ionized water [40].

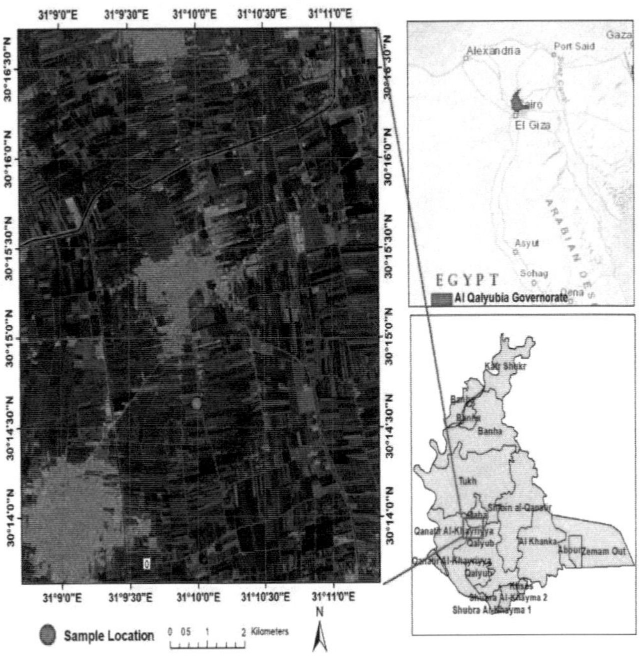

Figure 1. The site of study

Figure 2. Cabbage

Figure 3. Mallow

Figure 4. Turnip

3. Instrumentation

The samples were analyzed using Solar S4 atomic absorption spectrometer system which installed in the faculty of women for arts, science and education, Ain Shams University, Egypt. A Quad Line deuterium lamp is used in this AA Spectrometer for background correction, which has a double beam optical system. It has an Ebert monochromatic with a resolution of 1800 lines/mm. The system allows chemical examination of various materials, mostly metals and metal alloys, after the sample has been dissolved. Nearly most of the elements (with the exclusion of non-metals) can be determined in a broad variety of concentrations using an atomic absorption spectrometer (from 0.0001 to some dozen of percentage). The system is simple to operate and provides fast, precise and repeatable analysis results. The device is easy to use and provides fast, accurate and reproducible results of the analysis. The parameter conditions of the instrument are shown in 'table 1'.

Table (1): Operating Conditions for elements measured by FAAS

Element	Wavelength (nm)	Lamp Current (mA)	Burner Height (mm)	Fuel Flow (L/min)	Band Pass (nm)	Flame Type
Cu	324.2	5	7.0	0.9	0.5	Air-Acetylene
Mn	279.5	12	7.0	1.0	0.2	Air-Acetylene
Cd	228.8	8	3.8	0.8	0.5	Air-Acetylene

Pollution index

Using a pollution index, the degree of soil pollution caused by each heavy metal was calculated (PI), calculated as:

$$PI = \frac{C_c}{C_r}$$

where C_r and C_c represent the permissible heavy metal concentration and its estimated concentration, respectively [41].

Daily Intake of Metals (DIM)

The daily intake of metals (DIM) was estimated depending on the average consummation of contaminated-plants for both children and adults. It was evaluated as [42]:

$$DIM = \frac{C_{metal} \times C_{factor} \times D_{food\ intake}}{B_{average\ weight}}$$

where C_{metal} is the heavy metal concentrations in plants (mg/kg), C_{factor} is a conversion factor, $D_{food\ intake}$ represents the daily intake of vegetables, and $B_{average\ weight}$ represents the average body weight. A conversion factor of 0.085 was used to transform fresh green vegetable weight to dry weight [37]. The average daily intake for adults and children was assumed to be 0.345 and 0.232 kg per person per day, while the average adult and child body weights were assumed to be 55.9 and 32.7 kg [7,38]. These values were used for the calculation of HRI as well.

Health risk index (HRI)

The HRI was measured by dividing the normal consumption of metals in food by the oral reference dosage (RfD) using the following equation:

$$HRI = \frac{DIM}{RfD}$$

Oral reference doses were 0.04, 0.14 and 0.001 mg/kg/day for Cu, Mn and Cd respectively. [43]

The HRI value greater than 1 was considered as not safe for human health [44] and may pose health hazards for the consumers.

4. Results and Discussion

The amounts of heavy metals in environmental samples in Qaha are shown in 'Table 2'. From this table, the highest values of Mn concentrations in (cabbage, mallow and turnip) leaves are 119.25 ±5.96 ppm, 284.9 ±14.25 ppm and 106 ±5.3 ppm respectively. Whereas the lowest values of Mn concentrations found in cabbage stem, mallow roots and turnip roots are 87 ±4.35 ppm, 58.5 ±2.93 ppm and 53.75 ±2.69 ppm respectively. Thus, the mean values of Mn content in cabbage, mallow and turnip samples are 103.83 ±5.19 ppm, 142.72 ±7.14 ppm and 79.33 ±3.97 ppm respectively. The results of Mn content in vegetable samples are about three-folds higher than the normal level of Mn in plant 6.61 ppm 'Table 3'. This finding shows that the vegetable samples of the study area are polluted by Mn metal due to soil pollution such as pesticides and fertilizers. On other hand, the minimum concentrations of Cu in cabbage stem, mallow stem and turnip roots are 9.875 ±0.49 ppm, 16.33 ±0.82 ppm and 23.38 ±1.17 ppm respectively. The maximum concentrations of Cu in (cabbage, mallow and turnip) leaves are 46.58 ±2.33 ppm, 73 ±3.65 ppm and 32.25 ±1.61 ppm in respectively. The mean values of Cu in cabbage, mallow and turnip samples of the study area are 25.28 ±1.263 ppm, 35.26 ±1.76 ppm and 29.29 ±1.45 ppm respectively. Cu concentration in vegetable samples is about two times lower than the 73 ppm permissible value for Cu in plants 'Table 3'. This observation indicates that Cu metal has not contaminated the studied area's vegetable samples. The minimum concentrations of Cd in cabbage roots, mallow stem and turnip roots are 0.175 $\pm 8.75\times10^{-3}$ ppm, 0.175 $\pm 8.75\times10^{-3}$ ppm and 1.5 ±0.075 ppm respectively. The maximum concentrations of Cd in cabbage leaves, mallow leaves and turnip stem 2 ±0.1 ppm, 3.5 ±0.175 ppm and 2.48 ±0.124 ppm respectively. The average values of Cd in cabbage, mallow and turnip samples of the study area are 0.792 ±0.0396 ppm, 1.3 ±0.0649 ppm and 1.99 ±0.099 ppm respectively. These values are about threefold higher than the normal level of Cd in plant 0.2 ppm 'table 3'. This finding indicates that the research area's vegetable samples have been contaminated with Cd metal as a result of mining practices and a lack of environmental control.

According to 'Table 2', average Mn levels in vegetable samples from the research region are higher than Cu and Cd levels, so the levels of toxic elements in the current study's vegetable samples can be ordered in the following order: Cu > Mn > Cd. The use of chemical fertilizers and insecticides in farm fields is blamed for the contamination of radioactive metals in vegetable samples from agricultural areas.

'Table 2' shows that the soil has a high Mn content of around 4.8 ±0.24 ppm. Mn concentration of soil is about two times lower than the permissible limit of Mn in soil (437 ppm) 'Table 3'. This finding shows that the soil sample of the study area is not polluted by Mn metal. On other hand, the concentration of Cu in the soil is 2.4 ±0.12 ppm. Cu content in soil is approximately two times smaller than the permissible level of Cu in soil of 100 ppm 'Table 3'. This result indicates that the soil sample from the research area is not contaminated with Cu metal.

Table 2: Heavy metal concentrations (Mean ± SD) in vegetable samples and soil.

Sample	Cu	Mn	Cd
Cabbage root	19.38 ±0.969	105.25 ±5.26	0.175 ±8.75× 10^{-3}
Cabbage stem	9.875 ±0.49	87 ±4.35	0.2 ±0.01
Cabbage leaf	46.58 ±2.33	119.25 ±5.96	2 ±0.1
Mallow root	16.45 ±0.82	58.5 ±2.93	0.225 ±0.011
Mallow stem	16.33 ±0.82	84.75 ±4.24	0.175 ±8.75× 10^{-3}
Mallow leaf	73 ±3.65	284.9 ±14.25	3.5 ±0.175
Turnip root	23.38 ±1.17	53.75 ±2.69	1.5 ±0.075
Turnip stem	31.25 ±1.56	78.25 ±3.91	2.48 ±0.124
Turnip leaf	32.25 ±1.61	106 ±5.3	2 ±0.1
Soil	2.4 ±0.12	4.8 ±0.24	1.75 ±0.088

'Chart 1' illustrates the concentrations of the metals (Cu, Mn and Cd) in the cabbage sample. The concentrations of Mn are about 119.25 ppm, 105.25 ppm and 87 ppm in the leaves, roots and stems, respectively. While the concentrations of Cu are 46.58 ppm, 19.38 ppm and 9.875 ppm in the leaves, roots and stems, respectively. And, finally the concentrations of Cd are 2 ppm, 0.2 ppm and 0.175 ppm in the leaves, stems and roots, respectively.
From this chart we can conclude that the concentrations of Mn, Cu and Cd increased from the roots to the leaves with high concentrations in the leaves in the cabbage sample.

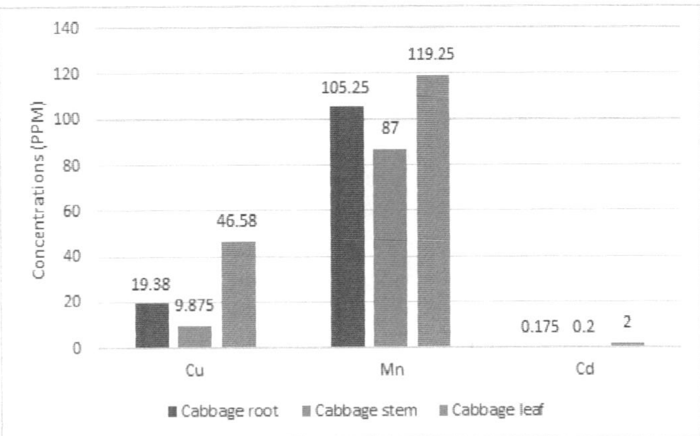

Chart 1: The concentrations of heavy metals in cabbage (root, stem, leaf)

'Chart 2' illustrates the concentrations of the metals (Cu, Mn and Cd) in the mallow sample. The concentrations of Mn are about 284.9 ppm, 84.75 ppm and 58.5 ppm in the leaves, stems and roots, respectively. While the concentrations of Cu are 73 ppm, 16.45 ppm and 16.33 ppm in the leaves, roots and stems, respectively. And, finally the concentrations of Cd are 3.5 ppm, 0.225 ppm and 0.175 ppm in the leaves, roots and stems, respectively.

From this chart we can conclude that the concentrations of Mn, Cu and Cd increased from the roots to the leaves with high concentrations in the leaves in the mallow sample.

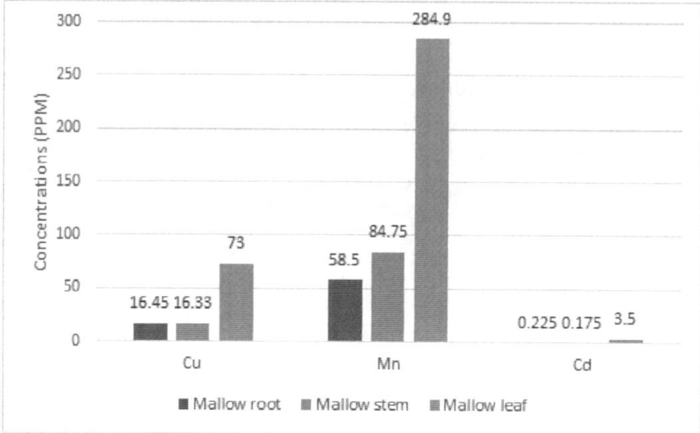

Chart 2: The concentrations of heavy metals in Mallow (root, stem, leaf)

'Chart 3' illustrates the concentrations of the metals (Cu, Mn and Cd) in the turnip sample. The concentrations of Mn are about 106 ppm, 78.25 ppm and 53.75 ppm in the leaves, stems and roots, respectively. While the concentrations of Cu are 32.25 ppm, 31.25 ppm and 23.38 ppm in the leaves, stems and roots, respectively. And, finally the concentrations of Cd are 2.48 ppm, 2 ppm and 1.5 ppm in the stems, leaves and roots, respectively.
From this chart we can conclude that the concentrations of Mn, Cu increased from the roots to the leaves with high concentrations in the leaves in the turnip sample. On the other hand, for Cd metal the high concentration was in the stems in the turnip sample.

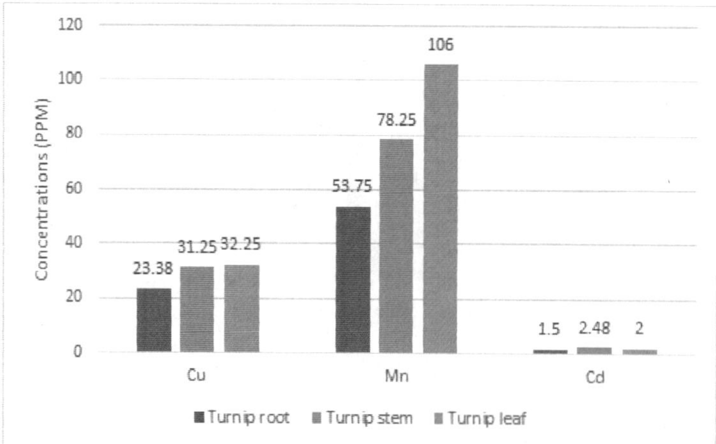

Chart 3: The concentrations of heavy elements in Turnip (root, stem, leaf)

Cd is present in the soil at a concentration of 1.75 ±0.088 ppm. This value is roughly two-thirds smaller than the appropriate Cd in soil cap of 3 ppm 'Table 3'. This finding indicates that the soil sample used in the analysis was not contaminated with Cd metal.

According to' Table 2', the average amount of Mn in the soil of the sample region is greater than the values of Cu and Cd, indicating that the toxic elements in the soil of the current study can be grouped in the following order: Mn > Cu > Cd.

'Chart 4' illustrates the concentrations of the metals (Cu, Mn and Cd) in the soil sample. The concentration of Mn is about 4.8 ppm. While the concentration of Cu is 2.4 ppm. And, finally the concentration of Cd is 1.75 ppm.

From this chart we can conclude that the soil has high concentrations of Mn, Cu and relatively low concentrations of Cd metal.

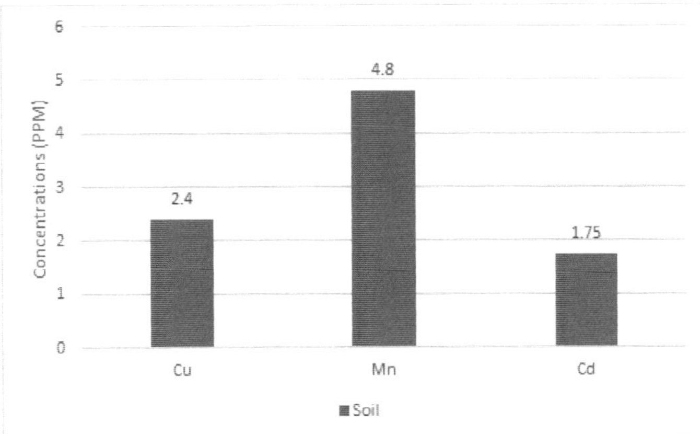

Chart 4: The concentrations of heavy metals in Soil

Table 3: Permissible concentration in agricultural systems (soil and plant) according to Food and agriculture organization (FAO)

Element	Permissible conc. In soil (ppm).(45)	Permissible conc. In plant (ppm).(46)
Cu	100	73
Mn	437	6.61
Cd	3	0.2

Pollution index

Soil is an important component of nature's dynamic structure, in which both chemical, biochemical, physical, and geological reactions occur with differing characteristics over time and space [47]. The physicochemical properties of the soil are undoubtedly influenced by the consistency of irrigation water. Irrigation of wastewater resulted in a significant improvement in soil pH and electric conductivity in the current sample. The migration, transition, and several reaction pathways of metal retention in soils are all influenced by pH, either directly or indirectly [48]. Furthermore, it is one of the most important factors affecting heavy metal mobility and adsorption in soils. Alghobar et al. (2014), in contrast to our findings, found that irrigation of waste water lowers soil pH; this may be attributed to the high content of organic matter, which decomposes to create organic acids.

Significantly, Mn, Cu and Cd had a low concentration in soil with pollution index less than the concentrations of these metals in vegetable samples are also high. This may be due to plants' high ability to absorb and transport heavy metals from the soil to the roots 'Table 4'.

According to 'Table 5', the current study's findings showed that Mn and Cd levels in vegetable samples surpassed the safe ranges, with an emission index greater than one. Cu, on the other hand, had an emissions index of less than one.

Table 4: The pollution index values in soil

Element	Pollution index values in soil
Cu	$0.024 \pm 1.2\times 10^{-3}$
Mn	$0.01 \pm 5.49\times 10^{-4}$
Cd	0.58 ±0.029

Table 5: The pollution index values in vegetable samples

Element	Pollution index values in cabbage	Pollution index values in Mallow	Pollution index values in Turnip
Cu	0.346 ±0.017	0.073 ±0.024	0.401 ±0.019
Mn	15.708 ±0.785	21.59 ±1.08	12 ±0.6
Cd	3.96 ±0.198	6.5 ±0.32	9.95 ±0.5

Daily intake of metals

The use of plants cultivated in wastewater-irrigated soils resulted in a high daily intake of metals (DIM) 'Table 6'. The DIM of all inspected heavy metals is lower than 1 for children and adults consuming cabbage, mallow and turnip plants.

Table (6): Daily Intake of Metals (mg/day) by adults and children for individual heavy metals in vegetable samples.

Heavy metal	Cabbage		Mallow		Turnip	
	Adult	Children	Adult	Children	Adult	Children
Cu	0.01	0.015	0.018	0.02	0.015	0.018
Mn	0.05	0.063	0.075	0.086	0.042	0.048
Cd	4.16×10^{-4}	4.77×10^{-4}	6.82×10^{-4}	7.84×10^{-4}	1.04×10^{-3}	1.2×10^{-3}

Health risk assessment

The HRI is a widely used method for determining the risk of hazardous materials in foods [7]. HRIs greater than 1 are thought to be harmful to human wellbeing. Cd had an HRI greater than 1 in the turnip sample irrigated with wastewater in this analysis. Cu and Mn had HRI of less than 1 for vegetable samples irrigated with wastewater as well.

Table (7): Health Risk Index (HRI) of heavy metals for adults and children via intake of heavy metals in vegetable samples.

Heavy metal	Cabbage		Mallow		Turnip	
	Adult	Children	Adult	Children	Adult	Children
Cu	0.25	0.375	0.45	0.5	0.375	0.45
Mn	0.357	0.45	0.54	0.61	0.3	0.34
Cd	0.416	0.477	0.682	0.784	1.04	1.2

5. Conclusion

The concentrations of heavy metals in environmental samples obtained from Qaha were studied using the FAAS technique. Due to human activity, chemicals, fertilizers, industrial activities, and the lack of environmental monitoring with pollution index greater than 1, the metals Mn and Cd are most significant in the vegetable samples of the study field. The findings of the heavy metals concentrations in this analysis indicate that the vegetable samples are not contaminated by Cu, with a pollution index of less than one. Heavy metal concentrations show that the soil sample is not contaminated by Cu, Mn, or Cd, with an emission index of less than one. Children and adults consuming cabbage, mallow and turnip plants have daily intake of metals of all inspected heavy metals lower than one. Furthermore, except for Cu and Mn metals, which had HRI less than 1, the Cd heavy metal suggested a high health risk from drinking turnip plant irrigated with wastewater. The observed results are important since human health is directly impacted by consuming vegetables. Monitoring of heavy metals in vegetables needs to be continued; because these are the main sources of food for human beings in the studied area and considered as bioindicators of environmental pollution.

References

[1] Elawa OE 2015. Impact assessment of industrial pollution on some economic plants south of Cairo Province, Egypt (Doctoral Dissertation, M. Sc. Thesis, Helwan Uni. Cairo, Egypt).

[2] Jolly YN, Islam A and Akbar S 2013. Transfer of metals from soil to vegetables and possible health risk assessment. *SpringerPlus*, **2**(1), p.385.

[3] Singh A, Sharma RK, Agrawal M and Marshall FM 2010. Risk assessment of heavy metal toxicity through contaminated vegetables from waste water irrigated area of Varanasi, India. *Tropical ecology*, **51**(2), pp.375-387.

[4] Harmanescu M, Alda LM, Bordean DM, Gogoasa I and Gergen I 2011. Heavy metals health risk assessment for population via consumption of vegetables grown in old mining area; a case study: Banat County, Romania. *Chemistry Central Journal*, **5**(1), p.64.

[5] Järup L 2003. Hazards of heavy metal contamination. *British medical bulletin*, **68**(1), pp.167-182.

[6] He G, Zhang L, Mol AP, Wang T and Lu Y 2014. Why small and medium chemical companies continue to pose severe environmental risks in rural China. *Environmental Pollution*, **185**, pp.158-167.

[7] Asgari K and Cornelis WM 2015. Heavy metal accumulation in soils and grains, and health risks associated with use of treated municipal wastewater in subsurface drip irrigation. *Environmental*

monitoring and assessment, **187**(7), p.410.

[8] Naggar YA, Naiem E, Mona M, Giesy JP and Seif A 2014. Metals in agricultural soils and plants in Egypt. *Toxicological & Environmental Chemistry*, **96**(5), pp.730-742.

[9] Chibuike GU and Obiora SC 2014. Heavy metal polluted soils: effect on plants and bioremediation methods. *Applied and environmental soil science*, 2014.

[10] Morgan R 2012. Soil, heavy metals, and human health. *In Soils and human health* (pp. 74-97). CRC Press.

[11] Al Naggar YA, Naiem ESA, Seif AI and Mona MH 2013. Honey bees and their products as a bio-in-dicator of environmental pollution with heavy metals. *Mellifera*, **13**, pp.1-20.

[12] Li XL, Gao J, Brierley G, Qiao YM, Zhang J and Yang YW 2013. Rangeland degradation on the Qinghai-Tibet plateau: Implications for rehabilitation. *Land degradation & development*, **24**(1), pp.72-80.

[13] Chen XW, Wong JTF, Mo WY, Man YB, Ng CWW and Wong M.H 2016. Ecological performance of the restored south east new territories (SENT) landfill in Hong Kong (2000–2012). *Land Degradation & Development*, **27**(6), pp.1664-1676.

[14] Ghoneim AM, Al-Zahrani S, El-Maghraby S and Al-Farraj A 2014. Heavy metal distribution in Fagonia indica and Cenchrus ciliaris native vegetation plant species. *J. Food Agric. Environ*, **12**, pp.320-324.

[15] Menazea AA, Ezzat HA, Omara W, Basyouni OH, Ibrahim SA, Mohamed AA, Tawfik W and Ibrahim MA 2020. Chitosan/graphene oxide composite as an effective removal of Ni, Cu, As, Cd and Pb from wastewater. *Computational and Theoretical Chemistry*, **1189**, p.112980.

[16] Ezzat H, Menazea AA, Omara W, Basyouni OH, Helmy SA, Mohamed AA, Tawfik W and Ibrahim M 2020. DFT: B3LYP/LANL2DZ study for the removal of Fe, Ni, Cu, As, Cd and Pb with Chitosan. *Biointerface Res. Appl. Chem*, **10**, pp.7002-7010.

[17] Farooq WA, Rasool KG, Tawfik W and Aldawood AS 2015. Application of laser induced breakdown spectroscopy in early detection of red palm weevil:(Rhynchophorus ferrugineus) infestation in date palm. *Plasma Science and Technology*, **17**(11), p.948.

[18] Tawfik W, Farooq WA, Al-Mutairi FN and Alahmed ZA 2015. Monitoring of Inorganic Elements in Desert Soil Using Laser-induced Breakdown Spectroscopy. *Lasers in Engineering (Old City Publishing)*, **32**.

[19] Iannelli MA, Pietrini F, Fiore L, Petrilli L and Massacci A 2002. Antioxidant response to cadmium in Phragmites australis plants. *Plant Physiology and Biochemistry*, **40**(11), pp.977-982.

[20] Bellés M, Rico A, Schuhmacher M, Domingo J and Corbella J 1995. Reduction of lead concentrations in vegetables grown in Tarragona Province, Spain, as a consequence of reduction of lead in gasoline. *Environment international*, **21**(6), pp.821-825.

[21] Schuhmacher M, Belles M, Rico A, Domingo JL and Corbella J 1996. Impact of reduction of lead in gasoline on the blood and hair lead levels in the population of Tarragona Province, Spain, 1990–1995. *Science of the Total Environment*, **184**(3), pp.203-209.

[22] Furman A and Laleli M 1999. Analysis of lead body burden in Turkey. *Science of the total environment*, **234**(1-3), pp.37-42.

[23] Palmgren F, Berkowicz R, Ziv A and Hertel O 1999. Actual car fleet emissions estimated from urban air quality measurements and street pollution models. *Science of the Total Environment*, **235**(1-3), pp.101-109.

[24] Calabrese EJ, Stanek E, James RC and Roberts SM 1997. Soil ingestion: a concern for acute toxicity in children. *Environmental health perspectives*, **105**(12), pp.1354-1358.

[25] Begerow J, Freier I, Turfeld M, Krämer U and Dunemann L 1994. Internal lead and cadmium exposure in 6-year-old children from western and eastern Germany. *International archives of occupational and environmental health*, **66**(4), pp.243-248.

[26] Farooq WA, Fatehmulla A, Yakuphanoglu F, Yahia IS, Ali SM, Atif M, Aslam M and Tawfik W 2014. Photovoltaic characteristics of solar cells based on nanostructured titanium dioxide

sensitized with fluorescein sodium salt. *Theoretical and Experimental Chemistry*, **50**(2), pp.121-126.

[27] Rao BH, Bhat AM and Singh DN 2007. Application of impedance spectroscopy for modeling flow of AC in soils. *Geomechanics and Geoengineering: An International Journal*, **2**(3), pp.197-206.

[28] Moenke-Blankenburg L, Schumann T and Nölte J 1994. Direct solid soil analysis by laser ablation inductively coupled plasma atomic emission spectrometry. *Journal of Analytical Atomic Spectrometry*, **9**(9), pp.1059-1062.

[29] Mohamed WT, Chen G, Kim J, Tao GX, Ahn J and Kim DE 2011. Controlling the length of plasma waveguide up to 5 mm, produced by femtosecond laser pulses in atomic clustered gas. *Optics express*, **19**(17), pp.15919-15928.

[30] Jarota A, Pastorczak E, Tawfik W, Xue B, Kania R, Abramczyk H and Kobayashi T 2019. Exploring the ultrafast dynamics of a diarylethene derivative using sub-10 fs laser pulses. *Physical Chemistry Chemical Physics*, **21**(1), pp.192-204.

[31] Tawfik W, Omar MM, Gamal YE and El Nadi L 2005, March. Ultrafast Moving Bubbles Initiated During The Propagation of Focused laser Pulses in water. In AIP Conference Proceedings *American Institute of Physics* (**Vol. 748**, No. 1, pp. 280-288).

[32] Ahmad K, Tawfik W, Farooq WA and Singh JP 2014. Analysis of alumina-based titanium carbide composites by laser-induced breakdown spectroscopy. *Applied Physics A*, **117(3)**, pp.1315-1322.

[33] Tawfik W, Bousiakou LG, Qindeel R, Farooq WA, Alonizan NH and Fatani AJ 2015. Trace analysis of heavy metals in groundwater samples using laser induced breakdown spectroscopy (LIBS). *Opt. Electron. Adv. Mater.*, **9**, pp.185-192.

[34] Farooq WA, Atif M, Tawfik W, Alsalhi MS, Alahmed ZA, Sarfraz M and Singh JP 2014. Study of bacterial samples using laser induced breakdown spectroscopy. *Plasma Science and Technology*, **16**(12), p.1141.

[35] Sawaf S and Tawfik W 2014. Analysis of heavy elements in water with high sensitivity using laser induced breakdown spectroscopy. *Optoelectron Adv Mater*, **8**, pp.414-417.

[36] Farooq WA, Tawfik W, Alahmed ZA, Ahmad K and Singh JP 2014. Role of purging gases in the analysis of polycarbonate with laser-induced breakdown spectroscopy. *Journal of Russian Laser Research*, **35**(3), pp.252-262.

[37] Tawfik W, Farooq WA, Al-Mutairi FN and Alahmed ZA 2015. Monitoring of Inorganic Elements in Desert Soil Using Laser-induced Breakdown Spectroscopy. *Lasers in Engineering (Old City Publishing)*, **32**.

[38] Mohamed WTY 2007. "Calibration free laser-induced breakdown spectroscopy (LIBS) identification of seawater salinity." *Opt. Appl* **37**, no. 1-2, 5-19.

[39] Maurya A, Kesharwani L and Mishra MK 2018. Analysis of Heavy Metal in Soil through Atomic Absorption Spectroscopy for Forensic Consideration. *Int. J. Res. Appl. Sci. Eng. Technol*, **6**, pp.231-242.

[40] Allan J, Mitchell T, Harborne N, Bohm L and Crane-Robinson C 1986. Roles of H1 domains in determining higher order chromatin structure and H1 location. *Journal of molecular biology*, **187**(4), pp.591-601.

[41] Liu WH, Zhao JZ, Ouyang ZY, Söderlund L and Liu GH 2005. Impacts of sewage irrigation on heavy metal distribution and contamination in Beijing, China. *Environment International*, **31**(6), pp.805-812.

[42] Khan S, Cao Q, Zheng YM, Huang YZ and Zhu YG 2008. Health risks of heavy metals in contaminated soils and food crops irrigated with wastewater in Beijing, China. *Environmental pollution*, **152**(3), pp.686-692.

[43] Rattan RK, Datta SP, Chhonkar PK, Suribabu K and Singh AK 2005. Long-term impact of irrigation with sewage effluents on heavy metal content in soils, crops and groundwater—a case study. *Agriculture, ecosystems & environment*, **109**(3-4), pp.310-322.

[44] Arora M, Kiran B, Rani S, Rani A, Kaur B and Mittal N 2008. Heavy metal accumulation in vegetables irrigated with water from different sources. *Food chemistry*, **111**(4), pp.811-815.

[45] USEPA (US Environmental Protection Agency) 2002. Region 9, Preliminary Remediation Goals. http://www.epa.gov/region09/waste/sfund/prg>.

[46] Kabata-Pendias A and Mukherjee AB 2007. Trace elements from soil to human. *Springer Science & Business Media.*

[47] Duressa TF and Leta S 2015. Determination of levels of As, Cd, Cr, Hg and Pb in soils and some vegetables taken from river mojo water irrigated farmland at Koka Village, Oromia State, East Ethiopia. *International Journal of Sciences: Basic and Applied Research*, **21**(2), pp.352-372.

[48] Zhao H, Xia B, Fan C, Zhao P and Shen S 2012. Human health risk from soil heavy metal contamination under different land uses near Dabaoshan Mine, Southern China. *Science of the Total Environment*, **417**, pp.45-54.

[49] Alghobar MA, Ramachandra L and Suresha S 2014. Effect of sewage water irrigation on soil properties and evaluation of the accumulation of elements in Grass crop in Mysore city, Karnataka, India. *American Journal of Environmental Protection*, **3**(5), pp.283-291.

Enhancement of the photovoltaic performance of dye-sensitized solar cell using porous silicon layer as photoelectrode

Y Gamal[1], T Abdallah[2]* and G M Youssef[2]

[1] October High Institute for Engineering & Technology, Cairo, Egypt.
[2] Physics Department, Faculty of Science, Ain Shams University, Abbassia, Cairo, Egypt.

*Corresponding author e-mail address: tamer_sharaf@sci.asu.edu.eg (T. Abdallah)

Abstract. Porous Silicon Dye-sensitized solar cell (PS-DSSC) with N719 Dye was employed as photoelectrode. PS layers were formed on textured crystalline silicon CZ-Si (100) by electrochemical etching (ECE) in hydrogen fluoride (HF) based electrolyte at constant current density for different etching times. The morphological properties of the PS were investigated by scanning electron microscopy (SEM). The optical properties of the textured surfaces are studied using photoluminescence (PL) and reflectivity measurements. The bandgaps of PS from UV-Vis and PL measurements increase to 1.9 eV. The Current-Voltage (I-V) characteristics show that the short-circuit current density J_{sc} and the open-circuit voltage V_{oc} increased as the porosity of the PS layer increased. These results show an improvement in the efficiency of PS-DSSC.
Keywords. Photovoltaics, Dye-sensitized solar cell, Porous Silicon, Electrochemical etching.

1. Introduction

Lately, photovoltaic devices, especially dye-sensitized solar cells (DSSCs), attract widespread academic and commercial attention by increasing the light-harvesting efficiency (LHE) of photovoltaic devices because of their low cost and higher efficiency [1-5]. Therefore, many approaches were made to improve the LHE, especially the improvement done through the photoanode. For example, plasmonic photoanodes, photonic crystal photoanodes, a porous active layer, hierarchically nanostructured photoanode, scattering layers [6-11].

One of the most critical issues is changing the optical design of the DSSC to improve the absorbance of light through the cell. Despite difficulties facing the researcher, such as losing the light reflectance and dye adsorption on the wide bandgap semiconductor. Scientists were hoping to discover new structures to enhance the light trapping through the photoanode. The texture of the surface plays a vital role in enhancing the conversion efficiency of the photovoltaic solar cell by lowering the reflection of light from the front surface, increasing the light trip inside the cell. As a result, the surface area of the photoanode increases[12-15]. The use of textured surface substrates as photoanode for DSSCs has been reported in some articles recently [16-17]. Thus, DSSC provides an economically and technically credible complementary concept to traditional PN junction photovoltaic devices.

In traditional photovoltaic systems, the semiconductor responsible for both the light absorption and charge carrier transport; on the other side, the two functions are separated here. Light is adsorbed to the surface of a wideband semiconductor, which is absorbed by a sensitizer. Charge separation created at the interface through photoinduced electrons is injected from the dye into the conduction band of the wideband semiconductor and then injected from the conduction band of the semiconductor to the charge collector. The role of dyes with a broad absorption band combining with oxide films of nanocrystalline morphology allows harvesting a large fraction of sunlight. Porous silicon (PS) has attracted much attention in the last years discovered in Bell Laboratories by accident as a byproduct of the wafer polishing process. This material is formed of a nanostructured PS skeleton. Porous silicon is well known for its abundance on the earth, good electrical conductivity, large specific surface area, and easy fabrication. PS is manufactured by electrochemical etching of silicon in hydrogen fluoride (HF) solution. In addition, the etchings methods help manage the morphological properties of PS layers (thickness and porosity) [18–23].

The critical factors in this method are the silicon doping type (P-type, n-type), the HF concentration, the electrolyte type (aqueous, organic, oxidant), the anodization etching time, and the surfactant. In this work, we use a porous silicon substrate as a photoanode in DSSC looked into the possibility of having a light-trapping process. First, we formed three different PS samples by using the electrochemical anodization etching process. Following that, ITO film was deposited on PSi substrates with electron beam sputtering, followed by the deposition of a wide bandgap (TiO_2) layer.

In this work, DSSCs have been constructed. An analysis involving SEM imaging for morphology, Photoluminescence spectroscopy (PL), UV–visible spectroscopy, and current-voltage measurements has been examined to characterize the proposed DSSC. Si substrates with different porosity layers changed both the dye absorption capacity and the optical property, making LHE far better.

2. Experimental

2.1 porous silicon fabrications

PS samples were fabricated by electrochemical etching of Czochralski (CZ) p-type [100] oriented silicon wafers with resistivity from (2 to 5) Ω. Cm and a thickness of 450 μm. firstly cut silicon samples into 1.5 × 1.5 cm^2 square samples. Before electrochemical etching, the wafers were and then dipped in trichloroethylene (isopropyl) and heated for 5 min then washed in de-ionized water, then dried under a stream of nitrogen (N_2) gas to remove all traces of native oxide [24-29]. Porous silicon was formed in a Teflon electrochemical cell with two electrodes silicon wafer as an anode and platinum contacts of the electrolyte as a cathode [30-35].

The PS was prepared by electrochemical anodization method in an electrolyte of 40 % hydrofluoric acid and 96 % ethanol [HF: C_2H_5OH = 2:1 by volume] at a constant current density of 20 mA/cm^2, and different etching time 15, 25, and 35 min. The samples were cleaned with acetone and then dipped in ethanol to remove the inorganic residuals, then used N_2 gas to dry the samples. Ethanol is always added to the solution to reduce the surface tension of HF, enabling the H_2 gas created during the reaction to avoid and prevent it from attaching to the etching surface and then enhancing the homogeneity of the resulting PS.

2.2 Preparation of photo-anode for DSSC

Firstly the PS substrate was cleaned with acetone and ethanol and then dried with N_2 gas. Next, the Indium Tin Oxide (ITO) was deposited on the PS layer using the sputtering technique where the system pressure was below 2×10^{-7} torr. After the growth, we annealed the samples at 400 oC under vacuum for 20 min. The thickness of ITO thin film is 50 nm. An area of 1x1 cm^2 was identified on a porous silicon substrate with a thin layer of ITO by using adhesive tape. The adhesive tape was used to define the deposition area and controls the film thickness. A drop of Tio_2 paste is spread by using a glass rode in activating area. The TiO_2 paste was prepared as follows: in a glass mortar, grind l gm of commercial Tio_2 powder (Degussa, P25) with 1 ml of distilled water containing (10 % v:v) acetylacetone (alpha, India), the acetylacetone is used as a dispersing agent. The mixture was then diluted by adding 1 ml of distilled water with continuous grinding. Reducing the suspension's surface tension is done by adding a few drops of triton X-100 (oxford, India) as a surfactant to disperse the TiO_2 nanoparticles in the colloidal. The obtained paste was transferred to a dropper bottle and stored in the dark until use. The paste was then dried on the samples at room temperature for 20 minutes and then sintered at 450 ^{o}c for about 30 minutes. After cooling, immersed the TiO_2 electrode in a $1.5x10^{-4}$M solution of the sensitizer dye, namely cis-bis (isothiocyanato) bis(2,2- bipyridyl-4,4-dicarboxylato) -ruthenium (II) (ruthenium-535, Solaronix) inanhydrous ethanol, for 24 h at room temperature, to absorb the dye on TiO_2. After that period, samples were rinsed with anhydrous ethanol to wash the unattached dye molecules away and dry.

A counter-electrode was prepared from platinum on an ITO glass substrate. The two electrodes were put together in a sandwich cell, and the electrolyte was in between them. Two electrolyte solutions were used triiodide/iodine (I^-/I^3) redox. The triiodide/iodine is prepared by mixing 0.1269 gm Iodine I_2 (0.05M) and 0.830 gm potassium iodide KI (0.5M) with 10 mL of ethylene glycol soluble them very well by continuous stirring.

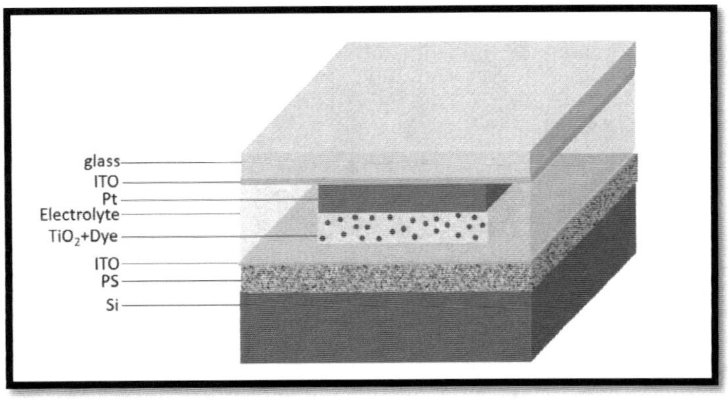

Figure 1 The schematic diagram for porous silicon dye-sensitized solar cell (PS-DSSC)

3. Results and Discussion

3.1 PS morphology of the surface

Figure 2 shows the surface morphology of porous silicon samples using a scanning electron microscope (SEM) during the etching process. Figure 2 (a) shows the surface morphology before PS formation, whereas the image shows many different sizes of spaced pyramids covered the surface. Figure 2 (b-d) shows the surface morphology after PS formation, which was fabricated with constant current density 20 mA/cm^2 and different etching times 15, 25, and 35 min at room temperature. The images illustrate the increase in surface area and porosities with different etching times.

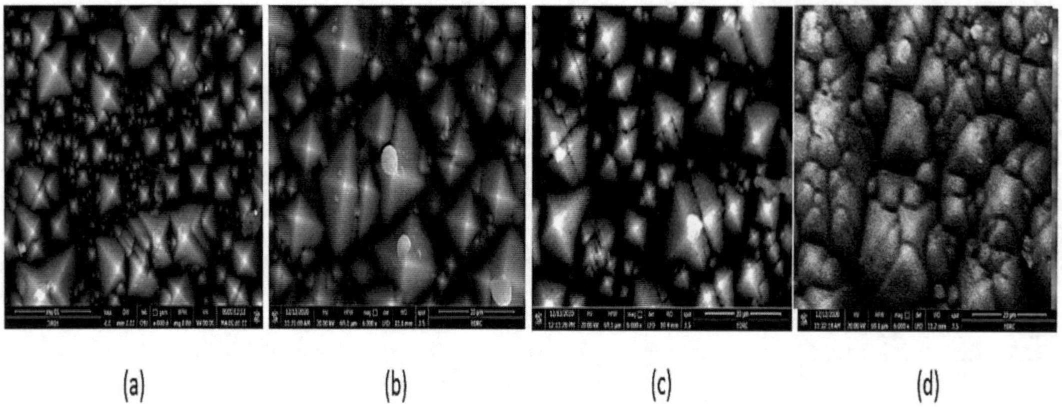

Figure 2 SEM image of top view (a) silicon substrate without etching (b) PS formed at 15 min etching time (c) PS formed at 25 min etching time (d) PS formed at 35 min etching time.

Figure (3) shows SEM images for the side view of the pyramids after depositing the ITO layer, which illustrates white light on the edge of pyramids, and TiO_2, which appears between and upon the pyramids. The high magnification shows the edge of the pyramid with a white beam of ITO and bubbles of TiO_2. In contrast, the ITO thickness is measured and found to be 49.71nm.

Figure 3 Side view of SEM image for ITO and TiO_2 on PS layer.

3.2 Optical properties

3.2.1 Photoluminescence (PL) of PS layers and pore size

Figure (4) show the photoluminescence spectra for PS samples with different etching times 15, 25, and 35 min and constant current density 20 mA/cm^2 with peaks at 655, 653.7, and 652.6 nm, respectively. The spectra show a blue shift indicates an increase in energy gap as etching time increased. The energy gap is calculated from

$$E_g = \frac{hc}{\lambda} = \frac{1240}{\lambda} \tag{1}$$

Where E_g is the Energy gap, h is Planck's constant, c is the speed of light and λ the peak wavelength of the photoluminescence. The average pore diameter was calculated [36]

$$E(ev) = E_g + \frac{h^2}{8d^2}\left[\frac{1}{m_e} + \frac{1}{m_h}\right] \tag{2}$$

The bandgap of bulk silicon Eg is 1.12 ev, h is Planck's contant, the pore diameter is d, m_e and m_h is the mass of electron and hole, respectively. substitution with E_g, h, m_h and m_e , this formula is simplified to

$$d = \sqrt{\frac{26.948}{E-1.12}} \tag{3}$$

The calculated pore diameter (d) is 5.90, 5.88, and 5.87 (nm) for 15, 25, and 35 min etching times. Thus, the pore diameter is decreasing with increasing the etching time.

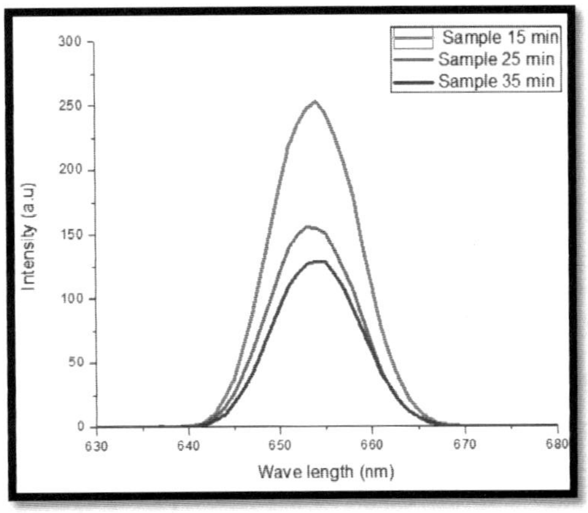

Figure 4 Photoluminescence spectra of porous silicon sample with different etching time

3.2.2 Reflectivity

Figure 5(a) shows the reflectance curve with different etching times. UV-Vis spectrophotometry was carried out to study optical changes on silicon after etching. We can calculate the energy bandgap by using the absorbance coefficient (α) and reflectance spectrograph (R) [37, 38].

$$\alpha = \frac{1}{2t} \ln \frac{R_{max} - R_{min}}{R - R_{min}} \tag{4}$$

IOP Conf. Series: Materials Science and Engineering 1171 (2021) 012002 doi:10.1088/1757-899X/1171/1/012002

The absorption coefficient is α that the film's thickness is t, R_{max}, and R_{min} are the maximum and minimum reflectance in the diffused reflection spectra. The reflectance for any intermediate photon energy is R. The calculated energy gap E_g computed from [39]

$$\alpha h\nu = A\,(h\nu - E_g)^m \tag{5}$$

The edge width parameter is A, m = ½ for direct transition, and 2 for indirect transition. As shown in figure 5(b), the graph was a plot between $(\alpha h\nu)^{½}$ and $h\nu$ then drawing the tangent line to determine the energy gap.

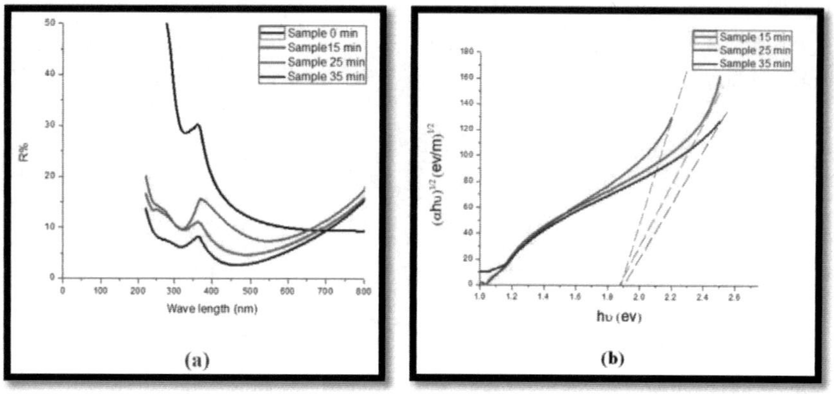

Figure 5 a) Reflectance curve b) The curve to determine the energy bandgap

Table 1 comparison between energy band gaps from UV-Vis spectrophotometry and photoluminescence

Etching time (min)	Energy band gap (eV)	
	UV	PL
0	1.100	1.12(literature)
15	1.880	1.893
25	1.885	1.897
35	1.900	1.900

3.3 I-V characteristic for PSDSSC

Figure (6) shows the I-V curve for PSDSSC with ruthenium dye for five etching times 0, 15, 20, 25, and 35 min. Table 2 shows the solar cell parameters. The curves show that the V_{OC} and J_{sc} increases to the highest open-circuit voltage values, 118 mV, and highest short-circuit current density of 0.223 mA/cm2 at 35min etching time. Moreover, the fill factor reaches the highest value of 0.42 and six times the conversion efficiency for the 35 min etching time.

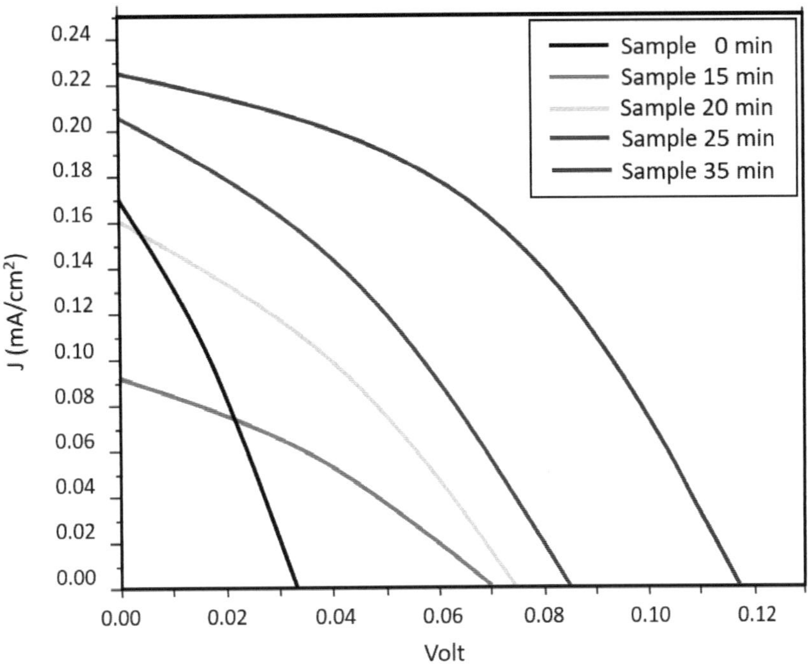

Figure 6 current-voltage characteristics of DSSCs fabricated by different etching time

Table (2) performance characteristics of DSSCs fabricated by different etching time

Time	V_{OC} (volt)	J_{SC} (mA/cm^2)	V_{max}(volt)	J_{max}(mA/cm^2)	P_{max}(μW/cm^2)	F.F
0	0.034	0.170	0.020	0.085	1.700	0.29
15	0.070	0.091	0.040	0.052	2.080	0.33
20	0.074	0.159	0.039	0.102	3.978	0.34
25	0.084	0.209	0.049	0.122	5.978	0.35
35	0.118	0.223	0.079	0.140	11.060	0.42

4. Conclusion

Porous silicon (PS) samples were prepared at constant etching current 20 mA/cm^2 at etching times 15, 20, 25, and 35 min used as photoanode in DSSC show that by increasing etching time, the energy bandgap was increased from 1.12 eV before etching to 1.9 eV after etching. The porosity increased, and the surface area of PS was increased, so we used Ps as photoelectrode in DSSC because it helped the dye solution to distribute over a large surface area and absorb much light, which improved the efficiency of DSSC. The best DSSC, which had PS with an etching time of 35 min their efficiency was increased six times more than silicon before any etching.

5. References

[1] Kim Y and Lee D 2011 Development of Dye-Sensitized Solar Cell for High Conversion Efficiency. Solar Cells *Dye-Sensitized Devices* https://doi.org/10.5772/19577.

[2] McCune M, Zhang W, and Deng Y. 2012 High Efficiency Dye-Sensitized Solar Cells Based on Three-Dimensional Multilayered ZnO Nanowire Arrays with "Caterpillar-like" Structure *Nano Letters* **12** 3656. https://doi.org/10.1021/nl301407b.

[3] Jung H S, Lee J K 2013 Dye Sensitized Solar Cells for Economically Viable Photovoltaic Systems. *The Journal of Physical Chemistry Letters*, **4** 1682. https://doi.org/10.1021/jz400112n.

[4] Adel R, Abdallah T, Moustafa Y M, Al-sabagh A M, and Talaat H 2015 Effect of polymer electrolyte on the performance of natural dye sensitized solar cells *Superlattices and Microstructures* **86** 62. https://doi.org/10.1016/j.spmi.2015.07.024.

[5] Adel R, Abdallah T, and Talaat H 2019 Efficiency enhancement of photovoltaic performance of dye sensitized solar cell using conducting polymer electrolyte of different functional group *Journal of Physics: Conference Series* **1253** 012029. https://doi.org/10.1088/1742-6596/1253/1/012029.

[6] Guo M, Chen J, Zhang J, Su H, Liu L, Fu N, and Xi K 2018 Coupling plasmonic nanoparticles with TiO2 nanotube photonic crystals for enhanced dye-sensitized solar cells performance. *Electrochimica Acta* **263** 373. https://doi.org/10.1016/j.electacta.2018.01.039.

[7] Shakeel A M, Pandy A K, and Abd Rahim N 2017 Advancements in the development of TiO2 photoanodes and its fabrication methods for dye sensitized solar cell (DSSC) applications. A review. *Renewable and Sustainable Energy Reviews*, **77** 89. https://doi.org/10.1016/j.rser.2017.03.129.

[8] Fan K, Yu J, and Ho W 2017 Improving photoanodes to obtain highly efficient dye-sensitized solar cells: a brief review. *Materials Horizons* **4** 319. https://doi.org/10.1039/c6mh00511j.

[9] Ghidelli M, Mascaretti L, Bricchi B R, Brognara A, Afifi T A, Russo V, Casari C S, and Bassi A L 2020 Light management in TiO2 thin films integrated with Au plasmonic nanoparticles. *Semiconductor Science and Technology* **35** 035016. https://doi.org/10.1088/1361-6641/ab6cea.

[10] Boro B, Gogoi B, Rajbongshi B M, and Ramchiary A 2018 Nanostructured TiO2/ZnO nanocomposite for dye-sensitized solar cells application: A review. *Renewable and Sustainable Energy Reviews* **81** 2264. https://doi.org/10.1016/j.rser.2017.06.035.

[11] Wang W, and Qi L 2019 Light Management with Patterned Micro- and Nanostructure Arrays for Photocatalysis, Photovoltaics, and Optoelectronic and Optical Devices. *Advanced Functional Materials* **29** 1807275. https://doi.org/10.1002/adfm.201807275.

[12] Aliaghayee M, Fard H G, and Zandi A 2016 Development of Macro-Porous Silicon Based Dye-Sensitized Solar Cells with Improved Light Trapping. *Journal of Electrochemical Science and Technology* **7** 218. https://doi.org/10.5229/jecst.2016.7.3.218.

[13] Boppella R, Mohammadpour A, Illa S, Farsinezhad S, Basak P, Shankar K, and Anorama S V 2016 Hierarchical rutile TiO2 aggregates: A high photonic strength material for optical and optoelectronic devices. *Acta Materialia* **119** 92. https://doi.org/10.1016/j.actamat.2016.08.004.

[14] Dye-sensitized solar cell fabrication and light trapping 2013 https://etda.libraries.psu.edu/files/final_submissions/8884.

[15] Elshimy H, Abdallah T, and Shama A A 2020 Optimization of Spin Coated TiO2 Layer for Hole-Free Perovskite Solar Cell. IOP Conference Series: *Materials Science and Engineering* **762** 012003. https://doi.org/10.1088/1757-899x/762/1/012003.

[16] Aliaghayee M, Ghafoori Fard H, and Zandi A 2015 A new method for improving the performance of dye sensitized solar cell using macro-porous silicon as photoanode. *Journal of Porous Materials* **22** 1617. https://doi.org/10.1007/s10934-015-0045-3.

[17] Aliaghayee M, Fard H G, and Zandi A 2016 Development of Macro-Porous Silicon Based Dye-Sensitized Solar Cells with Improved Light Trapping. *Journal of Electrochemical Science and Technology* **7** 218. https://doi.org/10.5229/jecst.2016.7.3.218.

[18] Han H, Huang Z, and Lee W 2014 Metal-assisted chemical etching of silicon and nanotechnology applications. *Nano Today* **9** 271. https://doi.org/10.1016/j.nantod.2014.04.013.

[19] Tontini G, Greaves M, Ghosh S, Bayram V, and Barg S 2020 MXene-based 3D porous

macrostructures for electrochemical energy storage. *Journal of Physics: Materials* **3** 022001. https://doi.org/10.1088/2515-7639/ab78f1.

[20] Salman K A, Omar K, and Hassan Z 2011 The effect of etching time of porous silicon on solar cell performance. *Superlattices and Microstructures* **50** 647. https://doi.org/10.1016/j.spmi.2011.09.006.

[21] Ramizy A, Aziz W J, Hassan Z, Omar K, and Ibrahim K 2011 Improved performance of solar cell based on porous silicon surfaces. *Optik* **122** 2075. https://doi.org/10.1016/j.ijleo.2010.11.026.

[22] Ramizy A, Aziz W J, Hassan Z, Omar K, and Ibrahim K 2010 The effect of porosity on the properties of silicon solar cell. *Microelectronics International* **27** 117. https://doi.org/10.1108/13565361011034812.

[23] Omar K, and Salman K A 2017 Effects of Electrochemical Etching Time on the Performance of Porous Silicon Solar Cells on Crystalline n-Type (100) and (111). *Journal of Nano Research* **46** 45. https://doi.org/10.4028/www.scientific.net/jnanor.46.45.

[24] Dantas M O S, Galeazzo E, Pees H E M, and Fernandez F J R 2002 Silicon micromechanical structures fabricated by electrochemical process. *Proceedings of IEEE Sensors*. https://doi.org/10.1109/icsens.2002.1037163.

[25] Theunissen M J, Appels J A, and Verkuylen W H 1970 Application of Preferential Electrochemical Etching of Silicon to Semiconductor Device Technology. *Journal of The Electrochemical Society* **117** 959. https://doi.org/10.1149/1.2407698.

[26] Lv H, Shen H, Jiang Y, Gao C, Zhao H, and Yuan J 2012 Porous-pyramids structured silicon surface with low reflectance over a broad band by electrochemical etching. *Applied Surface Science* **258** 5451. https://doi.org/10.1016/j.apsusc.2012.02.033

[27] Naderi N, and Hashim M R 2012 A combination of electroless and electrochemical etching methods for enhancing the uniformity of porous silicon substrate for light detection application. *Applied Surface Science* **258** 6436. https://doi.org/10.1016/j.apsusc.2012.03.056.

[28] Jakubowicz J 2003 Study of surface morphology of electrochemically etched n-Si (111) electrodes at different anodic potentials. *Crystal Research and Technology* **38** 313.

https://doi.org/10.1002/crat.200310037.

[29] Kolasinski K W 2005 Silicon nanostructures from electroless electrochemical etching. *Current Opinion in Solid State and Materials Science* **9** 73. https://doi.org/10.1016/j.cossms.2006.03.004.

[30] Hafez A A, Mohammed M M, Ibrahim I H, and Youssef G M 2021 An attempt to prepare an easy-fabricated porous silicon-based electrochemical non-enzymatic sensor for glucose detection. *Applied Physics A* **127**. https://doi.org/10.1007/s00339-021-04558-x.

[31] Youssef G M, El-Nahass M M, El-Zaiat S Y, and Farag M A 2015 Effect of porosity on the electrical and photoelectrical properties of textured n+ p silicon solar cells. *Materials Science in Semiconductor Processing* **39** 457. https://doi.org/10.1016/j.mssp.2015.05.054.

[32] El-Nahass M M, Youssef G M, Gad S A, and Noby S Z 2015. Electrical conduction mechanisms of CdTe quantum dots/p-si heterojunction. *Materials Science in Semiconductor Processing* **40** 337. https://doi.org/10.1016/j.mssp.2015.05.039.

[33] Youssef G M, El-Zaiat S Y, El-Malky M and Nawar H A 2015 Preparation and physical characterization of porous silicon layers for sensing applications. *International Journal of Applied, Physical and Bio-Chemistry Research (IJAPBCR).*

[34] Youssef G M, M. El-Nahss M, El-Zaiat S Y, and Farag M A 2015 Investigation of size and band gap distributions of nanoparticles from morphology and optical properties of porous silicon layers formed on a textured n+p silicon solar cell. *International Journal of Semiconductor Science & Technology (IJSST)*

[35] Ahmed A T, El Ghandoor H, El-Aasser M A, and Youssef G M 2019 Investigation of Porous Silicon Layers Properties Using Speckle Techniques for Photovoltaic Applications. *Silicon* **12** 1603. https://doi.org/10.1007/s12633-019-00255-w.

[36] El-Brolossy T A, Abdallah S, Abdallah T, Awad H, Mohamed M B, Negm S, and Talaat H 2008.

Photoacoustic spectroscopy characterization of CdSe quantum rods. *The European Physical Journal Special Topics* **153** 369. https://doi.org/10.1140/epjst/e2008-00464-x.
[37] El-Nahass M M 1992 Structural and electrical properties of cadmium-sulpho-selenide solid solutions. *Journal of Materials Science: Materials in Electronics*, **3** 71. https://doi.org/10.1007/bf00701097.
[38] Tauc J, and Menth A 1972 States in the gap. Journal of Non-Crystalline Solids, **8** 569. https://doi.org/10.1016/0022-3093(72)90194-9.
[39] Handbook of Photovoltaic Science and Engineering 2003 https://doi.org/10.1002/0470014008.

Acknowledgments

The generous support of the Egyptian STDF grant ID 4929 is greatly appreciated.

Laser interferometric investigations on some physical properties of Natural bio- gases

R Ghazy[1], G Moustafa[1], F El- Mekawey[1]and S.A. Abdelgawad[2]*

[1]Laser Laboratory, Physics Dept., Faculty of Science, Tanta University, Tanta, Egypt.
[2]*Faculty of Engineering MUST University, Cairo, Egypt.
E-mail: sayed.shama@must.edu.eg

Abstract. In this work a Mach- Zehnder interferometer system [MZIS] was constructed and illuminated by using [He-Ne] laser. And a gas flow system [GFS] was strictly designed and constructed that was connected to [MZIS], this gas flow system was controlled by both temperature range between 313 to 348 K and pressure range from 60-90 cm-Hg. The [MZIS] and [GFS] were linked and used for determining the refractive index of Methane [CH_4] and Nitrogen [N_2] gases as a function of pressure $n[p]_{T,}$ and temperature $n[T]_P$. The variety of refractive index of gases regarding the pressure factor at steady temperature $(dn/dp)_T$ was resolved. Likewise, the variety of refractive index of gases regarding to temperature at steady pressure factor $[dn/dT]_P$ was determined. By applying both Maxwell's and Clausius-Mostii equations, some physical parameters of Methane and Nitrogen gases, which related to refractive index, like as optical permittivity, dielectric susceptibility, Specific refractivity, Molar refractivity, and Molecular radius were estimated. In addition, the volume expansion values of the investigated gas samples were evaluated. Theoretically, by using Gaussian 09 program we had got some chemical parameters for Methane and Nitrogen.

1. Introduction

We described the main dielectric properties of Methane and Nitrogen and the relation between them with refractive index and their variation with pressure and temperature. We need to provide an accurate experiment, which give us the behavior of the gas at different temperatures and pressures. Some of these experiments show that the optical behavior of gases related to gas density. A modified Lorentz –Lorenz mode was used to provide theoretical values of the refractive indices of some gas hydrate at pressure between 2 and 15 Mpa, and temperature range from 0°c to 12°c by O. Bonnefoy *et al* [1]. Also M. Musso *et al* [2] measured the refractive index of liquid Sulphur dioxide SO_2 by using a Michelson interferometer.

K. Betzler *et al* [3] used a simple approach for detecting parallel sample refractive indices for shifting the pattern of interference when the sample is rotated in one arm of a Mach-Zehnder interferometer.

D.L. Khokhlov [4] considered the problem of the interference of a single particle in the two-path experiment by using the Mach-Zehnder interferometer with a single photon. Also, Rui Wang *et al* [5] used a Mach- Zehnder interferometer to measure a refractive index sensing structure by an integrated circuit with a hybrid silica polymer waveguide which have thermo-optic effects.

1.1 Characteristics and Uses of Methane gas

Methane is a chemical compound with the chemical form CH_4. It is the simplest alkane, [about 87% by volume] of natural gas and perhaps the most abundant organic substance on earth [6]. As a result, it has proven to be a desirable fuel. Steam reformation is used in the chemical industry to convert methane to synthesis gas, which is a mixture of carbon monoxide CO and hydrogen H2. Methane is non-toxic, but it is extremely flammable and can combine with air to generate explosive combinations. It is flammable only across a small concentration range of 5–15 percent in air [7]. Methane can be produced in a variety of ways, including biological, industrial, and geological. The geological path is separated into two categories: organic (thermally generated, or thermogenic) and inorganic (abiotic). Thermogenic methane forms from the separation of natural material at elevated temperatures. Methane is generated abiotically, that is, without biological activity, from inorganic molecules. Methane is manufactured via the Sabatier process on the industrial way, using hydrogenated carbon dioxide CO_2. Methanogens in the colon can use CO_2 and H_2 to produce methane as a byproduct in the biological route [8]. The generation of methane gas in 30-62percent of people by enteric bacteria has been based on several studies on the association between production of methane and intestinal problems such as constipation, bowel syndrome, diverticulosis and colon cancer [9]·

Methane's bioactivity is unquestionably not a new concept in medicine. Studies have shown that methane can decrease bowel movements by changing intestine neuromuscular capacities and decreasing peristaltic pace and increasing the abundance of pig ilea withdrawal [10].

Methane is classified as basically harmless by the United States' Occupational Safety and Hazards Administration. When it comes to employing this gas as a medicinal agent, this appears to be excellent news. Furthermore, because of its physical capacity to penetrate membranes and diffuse into organelles, methane has good diffusion characteristics [11]. The fundamental cause of many diseases is extreme Oxidative Stress which is a major source of devastating reactive oxygen species with the mitochondrial respiratory chain, but antioxidants have limited therapeutic effect, which may be attributable to their impermeability [12].

When methane enters the nucleus and mitochondria, nuclear DNA and mitochondrial protection suggest that methane has a defense effect against infections caused by oxidative pressure. There are two routes of transport that have been feasible for studies, including the inhalation of methane gas and injection with the high-salt saline, and Methane is able to shed cells and organs against irritation, oxidation, and apoptosis [13].

The use of inflammable gas and biogas to power the immediate use of methane for energy purposes can be considered. In gas power stations, for example, natural gas is used as fuel. In contrast to steam generating stations, steam boilers in gas generating stations are disregarded and so chemicals may be converted into electricity more efficiently. Moreover, 45% less carbon dioxide is emitted by gas power plants than by coal fired power stations [14]. Typical examples for usage of biogas are electric heat generation in modified gas boilers or electricity and heat simultaneous generation in linked units. Energy components, especially low-temperature cells, as Proton Exchange Membrane Fuel Cells ((PEMFC) are another important use of methane which are frames that endure for a long time in the tho80-100°C range [15].

The potential of methane for climate change and global warming has recently been addressed. Methane concentrations in the atmosphere were stable until around 100 years ago, when they began to grow. Methane has a global warming potential of 25 compared to CO_2 over a 100-year period, according to estimates from 1992 [although accepted statistics are likely an underestimate [16].

1.2 Characteristics and Uses of Nitrogen gas

Nitrogen is the major element of the atmosphere (78.03% by volume, 75.5% by weight). Nitrogen gas is harmless, noncorrosive, and nonflammable. It is colorless, odorless, and tasteless. Nitrogen is non-combustible and does not support life. It is inert unless heated to extremely high temperatures, at which point it forms nitrides with some of the more active metals like lithium and magnesium. It can also react with oxygen to generate nitrogen oxides, and it can combine with hydrogen to generate ammonia in the presence of catalysts [17].

Nitrogen is used as a plant fertilizer. Terrestrial plants absorb N, chiefly in the roots from soil, though N can be absorbed from the leaves when urea etc. is applied by a foliar spray. Plants are unable to utilize N2 on their own. Some leguminous and non-leguminous plants, on the other hand, use N2 through an interdependent relationship with bacteria in the soil, such as rhizobia or diazotrophic endophytes Mineralization of organic matter, breakdown of organic waste, and chemical fertilizers are all sources of inorganic nitrogen compounds. Ammonium [NH4+] and nitrate [NO3-] are the most common inorganic forms of nitrogen in soils, while organic N molecules like amino acids can also be absorbed [18].

In medical dentistry, nitrous oxide is the most commonly utilized inhalation general anesthetic, and it is also commonly utilized in emergency rooms and ambulatory surgery centers but, it should be combined with alternative inhalation and/or blood vessel agents in deep sedative/general anesthetic techniques. Nevertheless, as a one agent, it is quite safe for hesitant dental patients and it is excellent for giving little and moderate sedation [19].

Oxidation of specific materials is a result of usage of heat in the presence of oxygen. It can cause discoloration and the formation of a carbon layer on the cut edge, affecting the completed product's look or preventing any coating or paint finish from adhering to the oxidized surface. Nitrogen gas plays an important role as it prevents oxidization of the cut edge by permitting the laser to work in an oxygen free environment [20].

In particular types of laser cutting machines, nitrogen is equally used as 'cleansing gas,' which can be used to confirm the laser-beam guidance from the resonator to the cutting head that can in a way or other change the power or shape of the beam. The right nitrogen gas pressure, purity and flow may be achieved, with optimal cost savings and efficiency for customers depending on the capacity and thickness of the laser and on the type of material to be cut. The thickness of laser-cut material ranges from a few millimeters to more than twenty-five millimeters [21].

2. Interferometry

Interference was defined as the outcome of the superposition of two beams or waves at distinct sites that add constructively or destructively. It was quickly discovered that the two waves must have a constant phase difference between them, implying coherence, in order to form the observed pattern of interference fringes. If the phase difference among two waves is not constant, the degree of coherence must be determined that shows how distinctly visible is the interference pattern [22].

There are two typical methods for splitting a single beam of light into two beams. The beam is split in the first technique by passing it through apparatus set side by side; this method, known as division of wave front, is only useful for modest sources. The other method, known as division of amplitude, involves splitting the beam at one or more partially reflecting surfaces, where one part of the light is reflected and the other is transmitted [23].

Interferometers, which use large path differences between two interfering beams to produce interference, provide some very interesting and useful applications of the interference phenomenon as the determination of refractive index, refractive index distribution, refractive index variation with temperature, pressure, and refractive index variation with both temperature and pressure were investigated [24]-[29].

The Mach-Zehnder is a mainly simple device for showing interferences by amplitude division. A beam splitter divides into two parts and then recombines it with a second beam splitter, which then reflects the floor mirrors and finally emerges into a correct collection lens. The efficacy of an interferometer of the Mach-Zehnder will depend on the equivalent path length of the two beam routes before it is recombined so that the four reflecting surfaces are generally positioned around their centers in a parallelogram corner [30] as shown in the Fig.1.

The MZI system standard optical targets can be mentioned below:

1- The path difference can be recognized at zero in the system, so as to limit the number of bands and thereby maximize the intensity.

2-The system absorbs a small light which causes two beams to emerge around 50% of the light that reaches the final image plane.

3-With no disturbance in either beam, the direction and spacing of the fringes obtained in the source and final image planes, as well as the fringe number at a specific point in the fringe plane, are the only essential modifications.

For the above advantages of MZI, we constructed and used it in this work for measuring the refractive indices of biogases and their variation with both temperature and pressure.

3. Spectroscopy and quantum chemical calculation

Spectroscopy is the study of electromagnetic waves' interactions with matter. The structure and behavior of molecules of various sizes can be studied using infrared spectroscopy [IR]. IR spectroscopy is a low-cost, quick, and powerful technique that has long been a common tool among analytical chemists. Also, it uses in a wide variety of fields, ranging from pharmaceuticals applications, to food sciences to medicinal applications and even to entomology [31]-[34].

Ultraviolet-Visible spectroscopy examines changes in the molecular energy levels resulting from electron transfer from non-bonding orbitals. The π-electron system, conjugated unsaturation, aromatic compounds, and conjugated non-bonding electron systems are all commonly covered. This absorption spectroscopy uses electromagnetic radiations among 190nm to 800nm and is split into Ultraviolet (UV, 190-400nm) and visible (VIS, 400-800nm) regions [35].

Theoretical simulations of the researched molecules were carried out in order to determine some quantum chemical characteristics derived from our molecules' structure [36]. We used the density functional theory DFT (B3LYP) method with the 6-31G basis (d) for this purpose, and the calculations were done with Gaussian software 09W [37],[38].

The various molecular quantum parameters such as energy gap ΔE_{gap}, chemical potential μ, electronegativity χ, the chemical hardness η and the chemical softness σ are expressed by the following relations [39]-[43];

$$\Delta E_{gap} = E_{HOMO} - E_{LUMO} \quad (1)$$

$$\mu = \frac{E_{HOMO}+E_{LUMO}}{2} \quad (2)$$

$$\chi = -\frac{E_{HOMO}+E_{LUMO}}{2} \quad (3)$$

$$\eta = \frac{E_{LUMO}-E_{HOMO}}{2} \quad (4)$$

$$\sigma = \frac{1}{\eta} \quad (5)$$

Where E_{HOMO} is the energy of the highest occupied molecular orbital and E_{LUMO} is the energy of the lowest unoccupied molecular orbital.

3. Experimental Technique

The Helium-Neon laser with 632.8nm in wavelength is suitable for use as light source in our laser interferometer experiment. The gas cell was built in the form of two fixed cylindrical glass cells. As illustrated in Fig.1. The gas under investigation was filled in the inner cylinder, and the outer cylinder was utilized to heat the inner cylinder (gas cell). A calibrated thermocouple is fixed through them for measuring the interior temperature of the gas's cell. The outside cylinder we constructed for heating up the temperature of the gas under investigation that used as a container of heating liquid water surrounding the cell of the gas. The temperature of the gas might be measured. The accuracy of temperature change was about ± 1°C. The temperature range changed from 313 to 348 K.

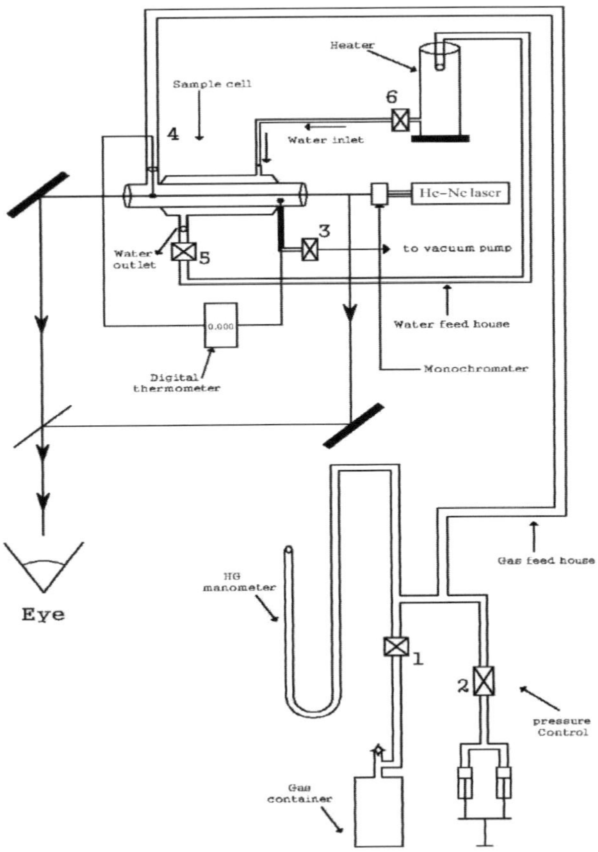

Figure 1. The constructed set- up used to measure the gas's refractive index.

We must change the pressure of the gas inside the cell to analyze the effect of pressure on the gas under investigation. In Fig.2, the change and control of gas pressure inside the cell is depicted schematically. The gas pressure inside the cell was controlled by two 100 mL syringes with a moveable piston controlled by two tiny screws. With an Hg-manometer, the amount of gas inside the gas cell was controlled.

The evacuation system was split into two sections (the rotary and the diffusion). The rotary has three ranges for pressure range the first from 1 to 10^{-1} mbar, the second from 10^{-1} to 10^{-2} mbar, and the third from $10^{-2}$to 10^{-3} mbar. After the rotary reached to the range of 10^{-3} mbar we could use the diffusion. The diffusion has three ranges for pressure, range one from 10^{-2} to 10^{-5} mbar, the second ranges from $2x10^{-5}$ to 10^{-6} and the third from $2x10^{-6}$ to 10^{-7}.

The number of fringes that traverse the field of view is changed by varying the gas pressure within the cell. By counting the number of changed fringes ΔN crossing the field of view by varying the pressure p in the gas cell, we can calculate the refractive index at any arbitrary pressure from the equation [44];

$$\boldsymbol{n(Ps)} = \frac{\lambda\, \Delta N}{t\, \Delta p}\, \boldsymbol{Ps} + \mathbf{1} \qquad \textbf{(6)}$$

Where; $n(Ps)$ is the refractive index at pressure Ps;(λ) is the wavelength of the laser light, (t) is the length of the cell, (ΔN) is the varying in the number of fringes count and(Δp), is the change in the pressure.

4. Theoretical Background

4.1. Permittivity and Dielectric Susceptibility

The value of permittivity ε is a dimensionless parameter it is one of the most important dielectric parameters, which plays a major role in the description of substance properties. In theory of the propagation of electromagnetic waves in a matter, the velocity of according to Maxwell's propagation is determined from the relation [45];

$$\boldsymbol{v} = \frac{1}{\sqrt{\varepsilon_\circ \varepsilon_r \mu_\circ \mu_r}} = \frac{c}{\sqrt{\varepsilon_r \mu_r}} \qquad \textbf{(7)}$$

Here, $\mu_r$is the relative permeability of medium, $\mu_\circ$ is the permeability in vacuum has value $4\pi \times 10^{-7}$ Hm^{-1}, ε_r is the relative permittivity in medium and $\varepsilon_\circ$ is the permittivity in vacuum has value 8.854×10^{-12} F/m. For the nonmagnetic dielectric materials, $\mu_r \approx 1$

$$\boldsymbol{v} = \frac{c}{\sqrt{\varepsilon_r}} \qquad \textbf{(8)}$$

Using equation (7) and (8), one got the relation between dielectric constant and the refractive index, assuming $\varepsilon_r = \varepsilon$

$$\boldsymbol{\varepsilon} = \boldsymbol{n}^{\mathbf{2}} \qquad \textbf{(9)}$$

In general, the dielectrics contain linear and non- linear dielectrics. In case of linear dielectrics, the relation between the value of polarization and the intensity of electric field [46] is given by:

$$\boldsymbol{P} = \boldsymbol{k_e}\, \boldsymbol{\varepsilon_0}\, \boldsymbol{E} \qquad \textbf{(10)}$$

Where k_e is the dielectric susceptibility and the product k_e ε_o, F/m, is known as the absolute dielectric susceptibility.

The quantity [47]

$$\boldsymbol{D} = \boldsymbol{\varepsilon_\circ}\, \boldsymbol{E} + \boldsymbol{P} \qquad \textbf{(11)}$$

Is called electric displacement (electric induction).

Also, there is a relationship between the electric displacement and intensity of an electric field

$$\boldsymbol{D} = \boldsymbol{\varepsilon\varepsilon_\circ}\, \boldsymbol{E} \qquad \textbf{(12)}$$

From equations (10), (11), and (12) we can get the relation between dielectric susceptibility and permittivity

$$\boldsymbol{k_e} = \boldsymbol{\varepsilon} - \mathbf{1} \qquad \textbf{(13)}$$

4.2. Specific and Molar Refractivity

From the Clausius- Mostti equation (46), (48)

$$\frac{\varepsilon-1}{\varepsilon+2}=\frac{N\alpha}{3\varepsilon_{\circ}} \quad (14)$$

Where, N= [N_AP/RT], here N_A is Avogadro's number, R is the gas universal constant and α is the polarizability. The quantity $\frac{\varepsilon-1}{\varepsilon+2}$ is known as specific refractivity [A_{sp}] of a gas

$$A_{SP}=\frac{\varepsilon-1}{\varepsilon+2} \quad (15)$$

By multiplying both sides of equation (15) by [$\frac{M}{\rho}$] we can get [49]

$$\frac{\varepsilon-1}{\varepsilon+2}\frac{M}{\rho}=\frac{N\alpha}{3\varepsilon_{\circ}}\frac{M}{\rho} \quad (16)$$

Where M is the molecular mass of a gas, ρ is the density of a gas, and $\frac{\varepsilon-1}{\varepsilon+2}\frac{M}{\rho}$ is known as molar refractivity **Π.**

4.3. Molecular Radius r and Volume Expansion β

When a conducting sphere of radius r is placed in an electric field E the induced moment P is determined by

$$P = r^3 E \quad (17)$$

But

$$P = \alpha E \quad (18)$$

From these two equations $\alpha = r^3$

From equation [14], by substituting [$\varepsilon_{\circ} = 1/4\pi$] we get

$$\frac{\varepsilon-1}{\varepsilon+2}=\frac{4\pi}{3}N\alpha \quad (19)$$

In case of cgs electrostatic units (without $4\pi\varepsilon_{\circ}$), equation (19) becomes [47]

$$\frac{\varepsilon-1}{\varepsilon+2}=\frac{4\pi}{3}N_A\alpha=\frac{4\pi}{3}N_A r^3 \quad (20)$$

There is a relation between the thermo- optic coefficient [dn/dT] and the volume expansivity β given by Murphy and Albert [50] as

$$\frac{dn}{dT}=\frac{-3n[n^2-1]}{2[2n^2+1]}\beta \quad (21)$$

5. Results and Discussion

5.1. Experimental calculations

We devoted to the experimental measurements of the refractive index and its variation with pressure and temperature for Methane and Nitrogen gases. Fig. 2.(a, b) implies that the refractive index of CH_4 and N_2 increase by increasing the pressure at constant temperature and selected wavelength 633nm. This is because by increasing of gas's pressure, the density of the gas increases; this leads the refractive index to increase. As shown in Table.1, the pressure coefficient values $(dn/dp)_T$ for methane and nitrogen decrease by increasing the temperature.

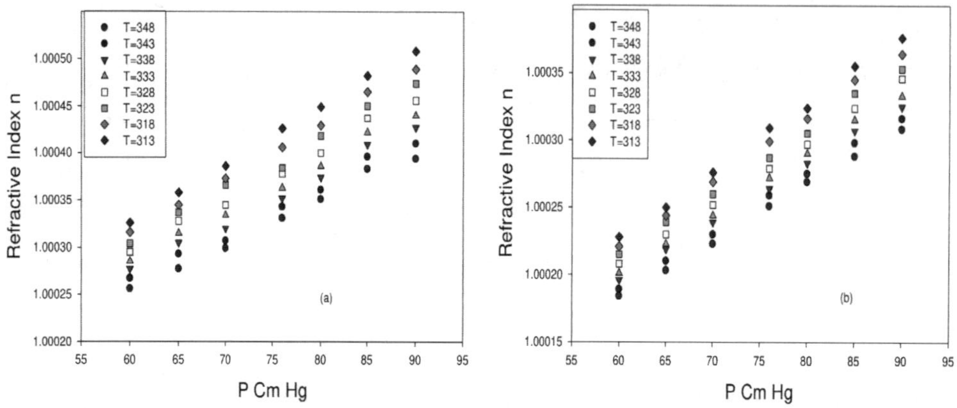

Figure 2. Refractive Index against Pressure for Methane CH_4 (a) and for Nitrogen N_2 (b) at constant Temperature.

Table 1. The values of pressure coefficient $(dn/dp)_T$ for methane and nitrogen gases at different temperature.

T K	348	343	338	333	328	323	318	313
$(dn/dp)_T$ $CH_4 \times 10^{-6}$	4.842	4.920	5.079	5.214	5.394	5.612	5.818	6.121
$(dn/dp)_T$ $N_2 \times 10^{-6}$	4.102	4.302	4.333	4.471	4.619	4.647	4.844	5.017

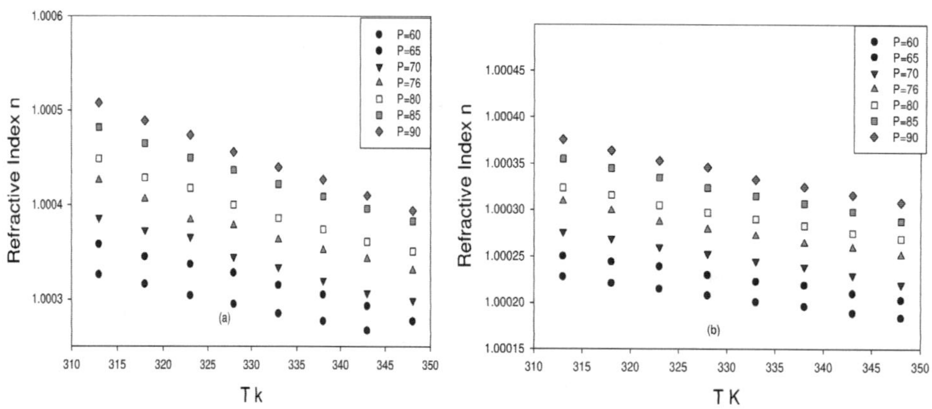

Figure 3. Refractive index against Temperature for Methane CH_4 (a) and for Nitrogen N_2 (b) at constant pressure.

Table 2. The values of the thermo-optical coefficient for methane and nitrogen gases at different pressures

PCm Hg	60	65	70	76	80	85	90
$(dn/dT)_p$ $CH_4 \times 10^{-6}$	-2.133	-2.228	-2.590	-2.597	-2.790	-2.800	-3.214
$(dn/dT)_p$ $N_2 \times 10^{-6}$	-1.266	-1.347	-1.566	-1.623	-1.748	-1.897	-1.935

From that figure the refractive index n decreases by increasing the temperature T at constant pressure p. This was reasonable because when the temperature of the gas increases, the density of the gas decreases, which leads to decrease the refractive index.

The values of the thermo-optical coefficient (dn/dT)p that shown in Table.2 describe how the gas can converge or diverge the incident laser beam according to the sign of (dn/dT) [51]. If the sign of the thermo-optical coefficient (dn/dT)p was positive, the gas converges the incident laser beam. If negative, the gas diverges the incident light. In our case the gases diverge the incident laser beam.

We calculated the refractive index n in the range of temperature 313-348 K and pressure from 60 to 90 Cm-Hg by using H_e-N_e laser with wavelength 633 nm. The calculated value of refractive index of Methane gas ranged from 1.000331 to 1.000426 from 348K to 313K at 76 Cm Hg or atmosphere pressure. Also, the Nitrogen gas has a refractive index n ranged from 1.000251 to 1.000309 from 348K to 313K at 76 Cm Hg as shown in Fig. 2, 3.

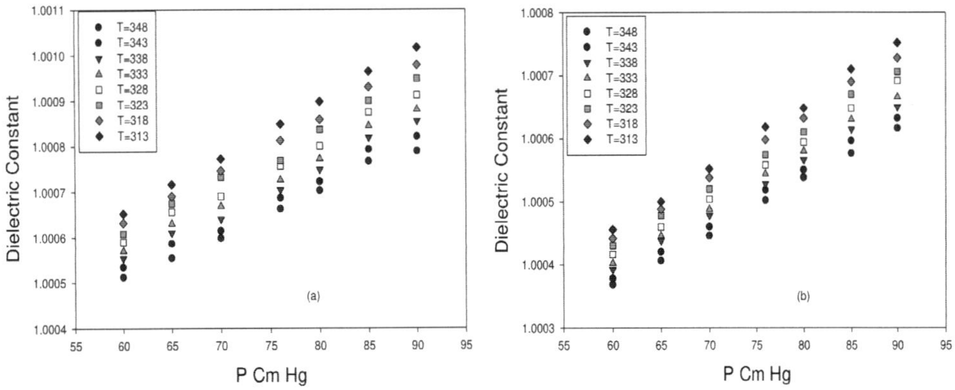

Figure 4. Dielectric Constant against Pressure for Methane CH_4 (a) and for Nitrogen N_2 (b) at constant Temperature.

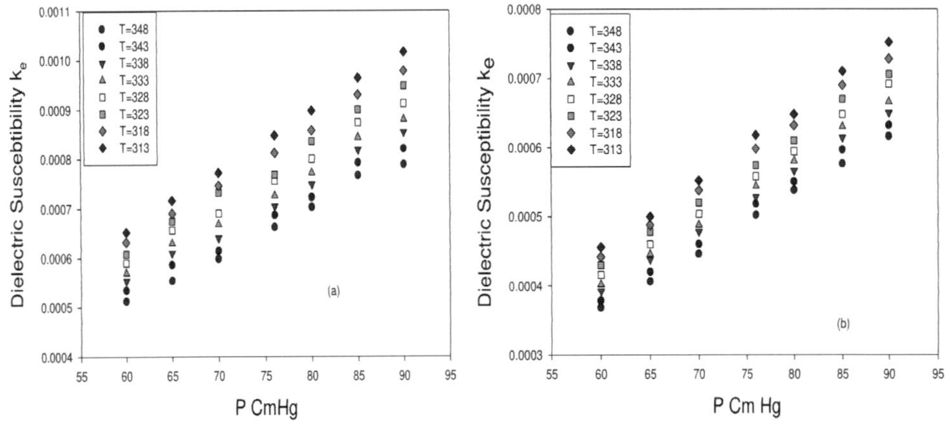

Figure 5. Dielectric Susceptibility against Pressure for Methane CH_4 (a) and for Nitrogen N_2 (b) at constant Temperature.

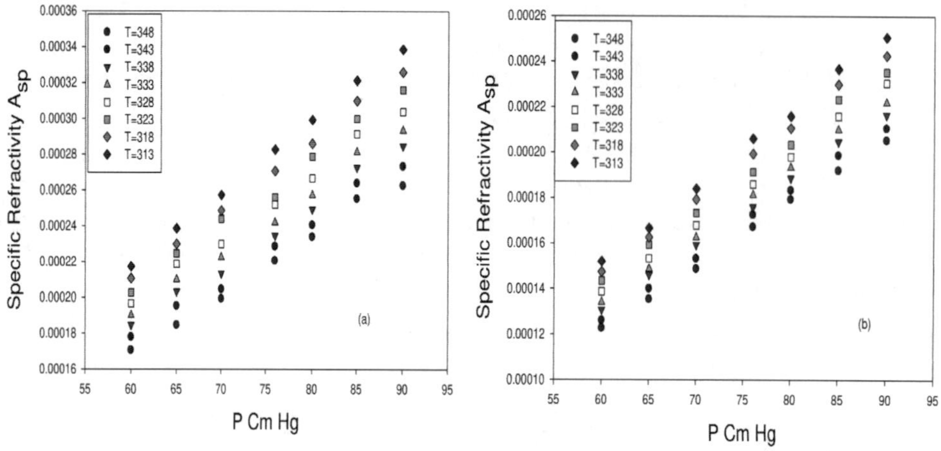

Figure 6. Specific refractivity against Pressure for Methane CH_4 (a) and for Nitrogen N_2 (b) at constant Temperature.

Fig.4,5 and 6 (a), (b) imply the relation between the dielectric constant ε , dielectric susceptibility K_e and specific refractivity A_{sp} of Methane and Nitrogen gases respectively with the pressure P at constant temperature T. These figures show that, by increasing the pressure of the gas at constant temperature, the permittivity, dielectric susceptibility, and specific refractivity increase. Tables 3, 4 show the calculated values of dielectric constant, dielectric susceptibility and specific refractivity of Methane, Nitrogen gases with respect pressure at constant temperature. These values decrease with increase the temperature.

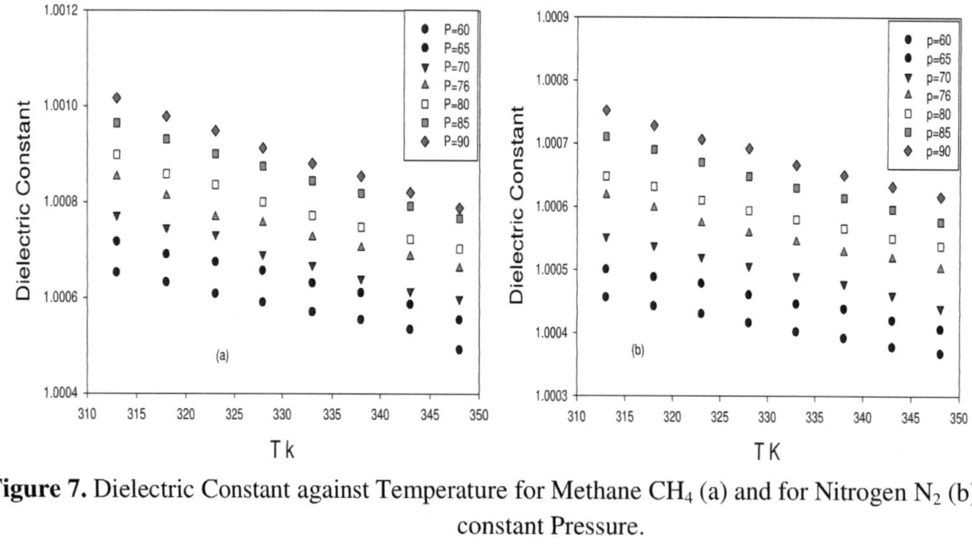

Figure 7. Dielectric Constant against Temperature for Methane CH_4 (a) and for Nitrogen N_2 (b) at constant Pressure.

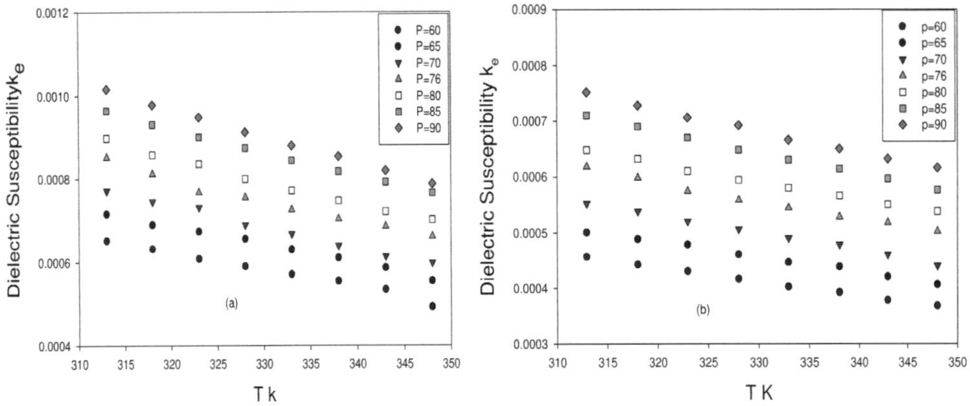

Figure 8. Dielectric Susceptibility against Temperature for Methane CH_4 (a) and for Nitrogen N_2 (b) at constant Pressure.

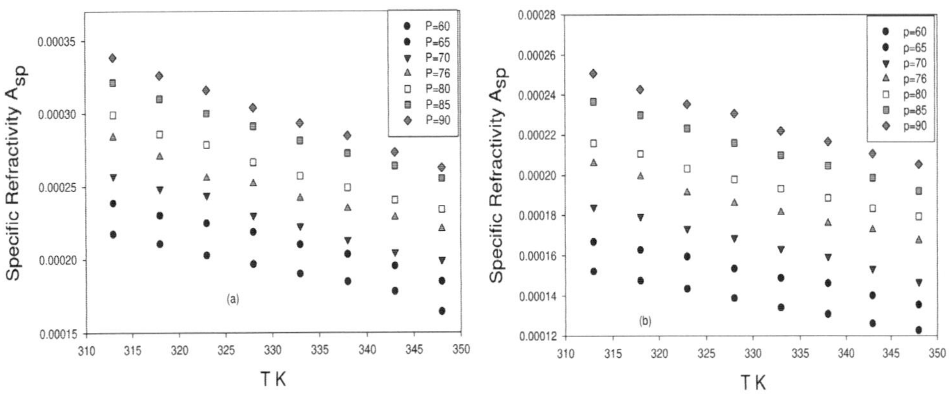

Figure 9. Specific Refractivity against Temperature for Methane CH_4 (a) and for Nitrogen N_2 (b) at constant Pressure.

Fig. 7,8 and 9 (a), (b) show the decreasing of dielectric constant ε, dielectric susceptibility k_e and specific refractivity A_{sp} by increasing the temperature of the gas at constant pressure for methane and nitrogen gases. Tables. 5, 6 show the calculated values of dielectric constant, dielectric susceptibility and specific refractivity of Methane and Nitrogen with respect to temperature at constant pressure. These values increase with increase the pressure.

Table 3. The values of $(d\varepsilon/dp)_T$, $(dk_e/dp)_T$ and $(dA_{SP}/dp)_T$ for Methane at different temperature.

T K	348	343	338	333	328	323	318	313
$(d\varepsilon/dp)_T$ $CH_4 \times 10^{-5}$	0.968	0.984	1.016	1.043	1.079	1.123	1.164	1.225
$(dk_e/dp)_T$ $CH_4 \times 10^{-5}$	0.968	0.984	1.016	1.043	1.079	1.123	1.164	1.225
$(dA_{SP}/dp)_T$ $CH_4 \times 10^{-6}$	3.228	3.279	3.386	3.476	3.596	3.741	3.878	4.080

Table 4. The values of $(d\varepsilon/dp)_T$, $(dk_e/dp)_T$ and $(dA_{SP}/dp)_T$ for Nitrogen at different temperature.

T K	348	343	338	333	328	323	318	313
$(d\varepsilon/dp)_T$ $N_2\times10^{-6}$	8.404	8.606	8.669	8.944	9.241	9.297	9.690	10.037
$(dk_e/dp)_T$ $N_2\times10^{-6}$	8.404	8.606	8.669	8.944	9.241	9.297	9.690	10.037
$(dA_{SP}/dp)_T$ $N_2\times 10^{-6}$	2.734	2.810	2.862	2.926	3.026	3.056	3.163	3.249

Table 5. The values of $(d\varepsilon/dT)_P$, $(dk_e/dT)_P$ and $(dA_{SP}/dT)_P$ for Methane at different pressure.

P Cm Hg	60	65	70	76	80	85	90
$(d\varepsilon/dT)_P$ $CH_4\times 10^{-6}$	-4.2678	-4.4586	-5.1827	-5.1971	5.5831	-5.6029	-6.4315
$(dk_e/dT)_P$ $CH_4\times 10^{-6}$	-4.2678	-4.4586	-5.1827	-5.1971	5.5831	-5.6029	-6.4315
$(dA_{sp}/dT)_P$ $CH_4\times 10^{-6}$	-1.4220	-1.4855	-1.7267	-1.7315	-1.8601	-1.8664	-2.1425

Table 6. The values of $(d\varepsilon/dT)_P$, $(dk_e/dT)_P$ and $(dA_{SP}/dT)_P$ for Nitrogen at different pressure

P Cm Hg	60	65	70	76	80	85	90
$(d\varepsilon/dT)_P$ $N_2\times 10^{-6}$	-2.533	-2.848	-3.134	-3.362	-3.558	-3.796	-3.872
$(dk_e/dT)_P$ $N_2\times 10^{-6}$	-2.533	-2.848	-3.134	-3.362	-3.558	-3.796	-3.872
$(dA_{sp}/dT)_P$ $N_2\times 10^{-6}$	-0.8443	-0.9491	-1.044	-1.120	-1.192	-1.264	-1.290

From equations (20), (21) we calculated the values of molecular radius and volume expansion of Methane, Nitrogen gases at P=76 Cm Hg, T= 313 as shown in table. 7.

Table 7. The values of molecular radius and volume expansion of Methane gas at P= 76 cm Hg and T= 313 K.

Gas	Molecular radius [r] A°	Volume expansion β
Methane [CH_4]	1.401	0.0060985
Nitrogen [N_2]	1.223	0.005255

5.2 Theoretical Calculations

All calculations were obtained with the Gaussian 09 program package. The structure optimizations were carried out at the DFT/6-31G(d) level of theory. The DFT method that was a significant improvement over Hartree-Fock is B3LYP. Generally, the B3LYP functional is faster than the most Post Hartree-Fock techniques and usually is a comparable method [52].

The CH_4 and N_2 molecular structures are optimized using DFT/B3YLB level theory and from which the HOMO (Highest Occupied Molecular Orbital) energy, LUMO (Lowest Unoccupied Molecular Orbital) energy and energy gap E_g are calculated as shown in Fig.10.

IOP Conf. Series: Materials Science and Engineering 1171 (2021) 012003 doi:10.1088/1757-899X/1171/1/012003

HOMO energy is often associated with the molecule's ability to give away its electrons to suitable vacant orbital, the low values of HOMO energy for Methane and Nitrogen indicated to difficulty of giving electrons to an acceptor having a vacant orbital molecular. On the other hand, LUMO gives information on the acceptance of electrons in the molecule. Methane and Nitrogen molecules have high value of LUMO energy which means disability to accept electrons from occupied orbital [52].

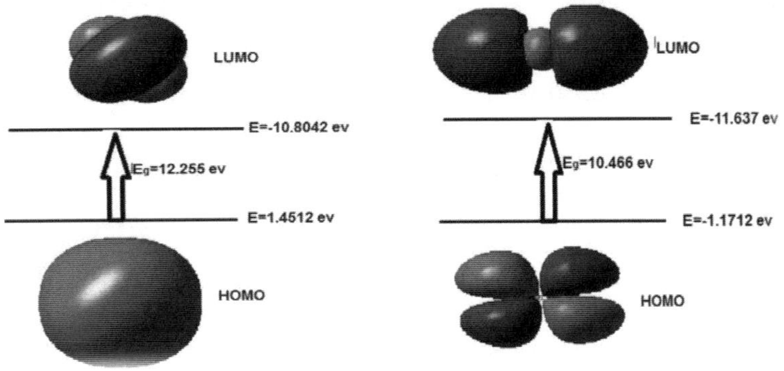

(a) (b)

Figure 10. The graphical presentation of the highest occupied HOMO and lowest unoccupied molecular LUMO orbitals for CH_4 (a) and N_2 (b).

Table. 8 Calculated E_{HOMO}, E_{LUMO} levels, energy gap E_g, dipole moment ρ and chemical hardness η, other quantum parameters chemical as electronegativity χ, chemical potential μ and chemical softness σ values of the studied Methane and Nitrogen molecules obtained by B3LYP/6-31G(d) level.

Parameter	Methane [CH_4]	Nitrogen [N_2]
E_{HOMO} [ev]	1.4512	-1.1712
E_{LUMO} [ev]	-10.8042	-11.637
E.G. [ev]	12.255	10.466
μ	-4.6765	-6.4044
H	-6.1277	-5.2331
P	0	0
X	4.6765	6.4044
σ	-0.1632	-0.1911
Molecular charge	0	0

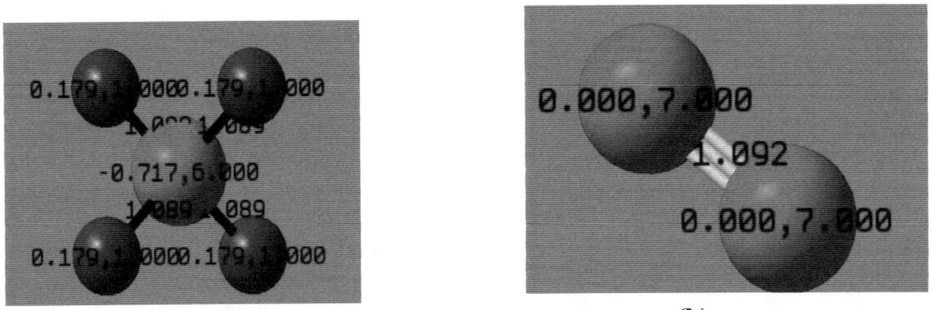

(a) **(b)**

Figure 11. Optimized structures of (a) Methane CH_4, (b) Nitrogen N_2 which have been obtained by using B3LYP/6-31G(d)ß level of theory.

The value of molecular charge equal zero which means the methane and nitrogen molecules is neutral as shown in Fig.11. The dipole moment ρ is the parameter most used to describe the polarity of a molecule [53]. It is defined as the charge product in the atom and the distance among two polar covalent bonds. Dipole moment of zero point to that there is no partial charge on either end of a covalent bond. On the other hand, chemical hardness η and softness σ are important chemical properties for measuring molecular stability and reactivity [52]; the results of these two parameters are also listed in Table.8. Methane and Nitrogen molecules have good chemical stability because of the increase in the hardness values and decrease in softness values.

In the infrared spectrum of methane there are two strong bands as shown in Fig.12, the fundamental frequencies, v_3 at 3020.3 cm^{-1}, and $v_4$at 1306.2 cm^{-1}. Because of the symmetry of the molecule, the other two fundamentals $v_1$at 2914.2 cm^{-1} and $v_2$at 1526 cm^{-1} are not active in the infrared, although these frequencies may appear in combination bands with the other frequencies and their overtones, to give bands in the near infrared [54].

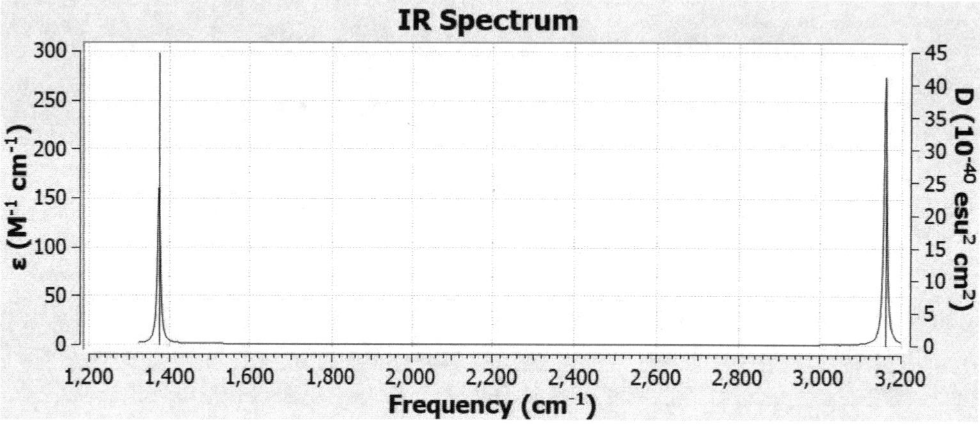

Figure 12. IR Spectrum of Methane which have been obtained by using B3LYP/6-31G(d) level of theory.

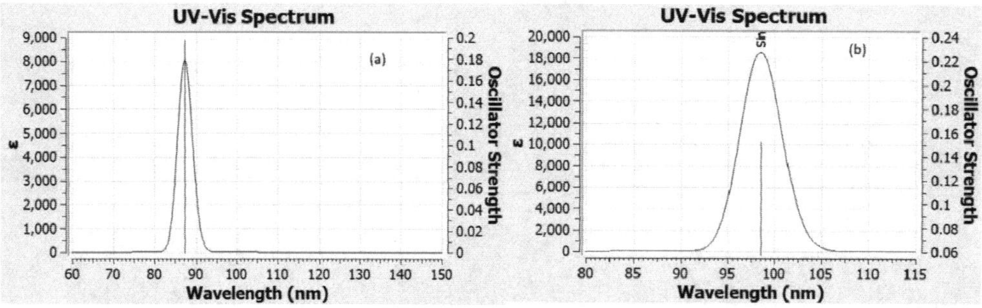

Figure 13. UV-V spectrum of Methane (a) and Nitrogen (b) which have been obtained by using B3LYP/6-31G(d) level of theory.

Chemical bonds are the forces that hold the atoms in a molecule together. The chemical forces that hold the atoms together are considered to be like to those utilized by massless springs. In a Cartesian coordinate system, each mass requires three coordinates to define the molecule's position in space, using coordinate axes x,y,z. Thus, the molecule has three independent degrees of freedom of movement. There will be a total of 3N degrees of freedom of movement for all the atoms in a molecule if a molecule has N atom. We get 3N-6 internal motions for a nonlinear molecule and 3N-5 for a linear molecule after subtracting the translational and rotational degrees of freedom from the 3N degrees of freedom. These interior vibrations are known as the normal modes of vibration. Consequently simple diatomic molecule has only one bond and only one vibrational band. If the molecule is symmetrical, e.g. N_2, the band is detected only in Raman spectrum not in IR spectrum. More intricate molecules with numerous bands have a more complicated vibrational spectrum, i.e. big molecules have several peaks in their IR spectrum [55]. By utilizing functional, B3LYP and applying the 6-31G [d] basis set, the UV-VIS spectra for Methane and Nitrogen gases were calculated. The calculated maximum absorption wavelength of Methane and Nitrogen molecules are 87, 98 nm respectively as shown in Fig.13.

6. Conclusion:

In the present text we had built an optical system to measure the refractive index of both Nitrogen and methane gases at different temperature and pressure. Some related physical properties were calculated. These calculated parameters are the optical permittivity (dielectric constant), dielectric susceptibility, specific refractivity, molar refractivity, molar radius volume expansion coefficient were estimated. In addition, both CH_4 and N_2 molecular structures are optimized using DFT/B3YLB level theory and from which the HOMO energy, LUMO energy and energy gap E_g were calculated with the Gaussian 09 program package. The structure optimizations were carried out at the DFT/6- 31G(d) level of theory. The DFT method that was a significant improvement over Hartree-Fock is B3LYP. Generally, the B3LYP functional is faster than the most Post Hartree-Fock techniques and usually is a comparable method. Methane and Nitrogen molecules have good chemical stability because of the increase in the hardness values and decrease in softness values. Also, the IR-spectrum of both methane and nitrogen gasses were investigated theoretically.

References

[1] Bonnefoy O, Gruy F and Herri J M 2005 *Materials Chemistry and Physics* **89**
[2] Musso M et al 2000 *Meas. Sci. Technol* **11**
[3] Betzler K et al 1988 *Rev. Sci. Instrum* **59** 652
[4] Khokhlov D L 2019 *Quant. Inf. Rev*. **7, No. 2**
[5] Wang R, Tang T, Shen J and Ch Li 2019 *Superlattices and Microstructures* **130**
[6] Boros M et al 2012 the anti- inflammatory effect of methane *Crit. Care. Med* **40**
[7] Ayhan D 2010 Gas Hydrate. In Methane Gas Hydrate, *New York Springer London Dordrecht Heidelberg*
[8] Roccarina D et al 2010 the role of methane in intestinal diseases *Am. J. Gastroenterol* **105**
[9] Sahakian A B, Jee S R and Pimental M 2010 Methane and the gastrointestinal tract *Dig. Dis. SC* **55**
[10] Jahng J et al 2011 The effects of methane and hydrogen gases produced by enteric bacteria on ileal motility and colonic transit time, *Neuro gastroenterol Motil*
[11] Pimental M et al 2006 *Methane, a gas produced by enteric bacteria, slows intestinal transit and augments small intestinal contractile activity*, *Am. J. Physical Gastrointest Liver Physo* **290**
[12] Smith R A and Murphy M P 2011 Mitochondria targeted antioxidants as therapies, *Discov Med* **11**
[13] Jia Y, Li Z, Liu C and Zhang J 2018 Methane medicine: A rising star gas with powerful anti-inflammation, antioxidant, and antiapoptosis properties, *Oxidative Medicine and Cellular*

Longevity

[14] Kwinta W 2010 the role of Methane in energy will increase, *Polska Energia- Polish- Energy* **1**

[15] Skorek A and Wlodarvzyk R 2018 the use of Methane in practical solutions of environmental engineering, *Journal of Ecological Engineering* **19(2)**

[16] Mohajan H 2011 dangerous effects of Methane gas in atmosphere, *Journal of Economic and Political Integration* **2(1)**

[17] Maxwell G 2004 Synthetic nitrogen products: A practical guide to the products and processes, *Springer Science & Business Media*

[18] Ohyama T 2010Nitrogen as a major essential element of plants, *Nitrogen. Assim. plants* **37(2)**

[19] Brown S M and Sneyd J R 2015 Nitrous oxide in modern anesthetic practice. *Journal of BJA Education*

[20] Eristman J W and Fowler D 2003 oxidized and reduced nitrogen in the atmosphere, *Eoless Publishers, Oxford, UK*

[21] Riveiro A, Quintero F and pou J 2010 laser fusion cutting of difficult materials, *Woodhead Publishing*

[22] Qureshi T 2019 Coherence, interference and visibility, *Arxiv Preprint Arixiv*, **1905** 00917

[23] Born M and Wolf E 1993 Principles of optics 6ed *Cambridge Univ. pess*

[24] Zhao G, Li F and Zhao C 2020 *Atmospheric Environmental* **240** 117800

[25] Wilmooth D M and Sayres D S 2020 *Journal of Quantitative Spectroscopy& Radiative Transfer* **255** 107224

[26] Richard E, Gaskin and Colin Lewise 1980, *Optical Acta* **27** 1287

[27] Daniel D, Young, Kenneth C 1972 *Jungling, Tomi, Williamson and Elgenen, R., Nichols, IEEE* **720**

[28] Hawkes J B and Astheimer R W 1948, *J. Opt. Soc. Am* **38** 617

[29] Vedam K, and Pichet Limsuwan 1977 *Rev. Sci. Instrum* **48** 245

[30] Zetie K P, Adams S F and Tocknell R M 2000 How does a Mach- Zehnder interferometer work? *Phys. Educ* **35(1)**

[31] Raggo Y et al 2006 A review of near infrared spectroscopy and chemometrics in pharmaceutical technologies, *J. Pharm. Biomed. Anal* **44**

[32] Wadood S A et al 2020 Recent development in the application of analytical techniques for the traceability and authenticity of food of plant origin. *Microchem. J.* **152** 104295

[33] Belskaya L V 2019 Use of IR spectroscopy in cancer Diagnosis *J. Appl. Spectrosc* **86**

[34] Johnson J B and Naiker M 2019 A review of the use of near- infrared spectroscopy NIRS in entomology. *Appl. Spectrosc. Rev* **55**

[35] Kumar S 2008 Spectroscopy of Organic Compounds

[36] El Aoufir Y et al 2016 Quinoxaline derivatives as corrosion inhibitors of carbon steel in hydrochloridric acid media: electrochemical, DFT and Monte Carlo simulations studies, *J. Mater. Environ. Sci.* **7(210)**

[37] Petersson A et al 1988 A complete basis set model chemistry and the total energies of closed-shell atoms and hydrides of the first- row elements *J. Chem. Phys.* **89(4)**

[38] Ansari K R and Quraishi M A 2015 Experimental and quantum chemical evaluation of Schiff bases of isatin as new and green corrosion inhibitors for mild steel in 20% H_2SO_4, *J. Taiwan. Inst. Chem. Eng* **54**

[39] Neese F 2012 An Ab ini tio, DFT and semiempirical SCF-MO package. *Information Technology Services, The University of Hong Kong*

[40] Becke A D 1986 Density functional calculations of molecular bond energies *J. Chem. Phys.* **84(8)**

[41] Lee C, Yang W and Parr R G 1988 Development of the Colle–Salvetti correlation-energy

formula into a functional of the electron density *Phys. Rev B.* **37(2)**
[42] Saha S K et al, 2014 Molecular dynamics and density functional theory study on corrosion inhibitory action of three substituted pyrazine derivatives on steel surface *Can. Chem. Trans.* **2(4)**
[43] Saha S K et al, 2015 Density functional theory and molecular dynamics simulation study on corrosion inhibition performance of mild steel by mercaptoquinoline Schiff base corrosion inhibitor *Phys E* **66**
[44] Schroedter M 2000 *Mathematica for Student, England*
[45] Atkins P W 1978 *Physical Chemistry English Longnage Bok Society and Oxford University Press, Oxford, ch* **23**
[46] Williams J H 2018 *the Molecules as Meme Morgan&Claypool Publishers*
[47] Tareef B 1979 Physics of Dielectric Materials Mir Puplishers**, *Moscow. Second ed***
[48] Huang X, Zhang H, Lai Y, Wang G, Li M, Hang C and Li J 2018 *European Journal of Inorganic Chemistry,* **17**
[49] Tuli G D and P.L. Soni P L 1969 *B. Sc. Physical Chemistry Pass-Book, Tenth edition*
[50] Murphy C and Alpert S 1971 *Am. J. Phys,* **39** 834
[51] Tarasev L V 1983 Laser Physics *Mir Publishers, Moscow*
[52] Fergachi O et al 2019 Corrosion inhibition of ordinary steel in 5.0 M HCl medium by benzimidazole derivatives: electrochemical, UV–visible spectrometry and DFT calculations *Journal of Bio-and Tribo-Corrosion* **5(1)**
[53] Wang H et al 2007 DFT study of new bipyrazole derivatives and their potential activity as corrosion inhibitors *J. Mol. Model* **13(1)**
[54] Nelson R C, Plyler E K and Benedict W S 1948 Absorption spectra of methane in the near infrared *J. Res. Natl. Bur. Stand* **41** (1934), 615
[55] Theophile T 2012 Infrared spectroscopy: Materials science*, engineering and technology BoD–Books on Demand*

IOP Conf. Series: Materials Science and Engineering 1171 (2021) 012004 doi:10.1088/1757-899X/1171/1/012004

The Reliable Concrete Compression Strength Assessment by SCHMIDT Hammer for Different Concrete Grades

Mostafa Abd Elrazek[1],Yasser Abdel Shafy Gamal[2]
[1] Lectures of properties and strength of materials, Civil Department, High Institute of Engineering Technology, EL-MINA.
[2] Lectures of structural engineering, Civil Department, High Institute of Engineering Technology, EL-MINA.

E-mail: (dr.mostafa.abd.k@mhiet.edu.eg)[1] , (Yasser.gamal2310@mhiet.edu.eg)[2]

Abstract. The evaluation of the existing construction building is one of the urgent problems in the current era, different tests and methodologies can be used to get the required information about the properties of the structure so that the construction industry interests in applying the concrete test with the lowest time and cost, The purpose of this study is to present the combination between Nondestructive test (NDT) and Machine compression Test (MCT) to get the real results for the Schmidt hammer test method. All tests are created for different grade of RC concrete from medium to high grade with classes between Grade 200 to Grade 800 , to study the Nondestructive test (NDT) and generate the comparison with the Machine compression Test (MCT) , using different composition of the concrete with Varity in the proportion mix between ingredients of cement , water content , aggregate, all results have been represented tabular and graphically to notice any deviation between Destructive test (DT) and MCT test, Regression analysis was been used as a statistical method to get the variation between tests and establish as a linear mathematical relationship between compressive strength and rebound number and it can be illustrated that the variances between them in case of Low. the medium and semi-high grade is very low which means the null hypothesis is accepted according to the statistical analysis, and on another side, there are variances in case of very high grade according to statistical analysis.

1. Introduction

In the reinforced concrete structure, Several parameters such as the aggregate Type, cement types, RC design, admixture design, and production methodologies can affect the reliability of RC construction building on resisting natural disasters (Yılmaz & Avsar, 2013)[1]. The lowest quality of concrete has always been one of the critical problems for the civil Engineers so that it is crucial to take all required precaution in the structure design, on the another hand the constructors always search the save way for evaluation the RC structure building with lowest time and cost, it can be obtained by using NDS (Nondestructive test), Schmidt hammer and Ultrasonic Pulse Velocity (UPV) which can be consider the famous test to save energy and cost.

The objective of studying is to study Nondestructive test reliability for the concrete compression strength by creating the relationship between NDT and MCT which means calibration curve for improving the estimation of the scores, to obtain these results, using two groups for the calibration curve, the first group of specimens was gotten from Schmidt Hammer curve, and the other group was created from the MCT test

The study investigates the creation of correlation curve and the difference between the rebound curve results and the actual reading from MCT and applies all tests for the different grade cases to get full comprehensive description about the deviation between them.

2. Literature Review

The concrete can be considered as a composite material with a combination of aggregate, water, and cement with different proportion, one of the important advantages of concrete is the ability to create with the desired size and shape with different concrete compressive strength, civil engineers take care about the ability of the concrete element to resist all anticipated load, therefore the compressive strength test should be checked to ensure the suitability of the building for living during the life cycle of the building, furthermore, the increase of the concrete age around the world led to numerous requirements for growing up the assessment of the concrete strength, so the concrete compressive strength is the most important characteristic to determine the quality of concrete (Abrams, 1927)[2].

In most cases, the concrete compressive strength should be determined with the samples during the pouring, but the advancement in the construction industry requires to develop of different tests to identify the strength of concrete elements quickly and applied these in place which became more urgent, and application of the tests on the concrete compressive strength became more important especially in the countries located in the seismic zone (Yon et al., 2015) [3]. However, a lot of existing building in the current era needs to determine the strength of concrete in a short time, So there are exigent needs for measuring the strength of the concrete in the site, which is based on the quality assurance for assessment of the concrete elements, NDT is the best economical and practical test to use and very popular due to permit the testing in place and evaluate theses with short time (Alyamac, 2017)[4].

Although the simplicity of the Schmidt hammer test, there are various factors that can be lead to the deviation results, such as Age of concrete, Roughness of concrete surface, and the properties of aggregate and cement have harmful effects on the results of the testing (Malhotra,2004)[5], so the testing the cleaning of RC surface should be applied, and for test recommendation, Ilhan suggested a recommendation for Schmidt test by ten readings should be done with distance not more than 20 mm and 40 -50 mm from the edge of the samples (Ilhan, 2000)[6].

The rehabilitation of the concrete building is related to the properties such as strength and durability (Hobbs et al, 2007)[7] so civil engineers interest to apply different tests to get the full information, in general, the determination of these properties needs two main types of test DT which can be executed by the drilling core test and another type is NDT on-site, the rebound hammer is one of type of NDT and it is a popular using by civil engineer due to simplicity and faster test.

The destructive test method (DT) can be executed by crushing the cast specimen until failure point, on the other side the Nondestructive test has an advantage in an application, it is carried out without destructive of the structural element with the lowest cost and time but it suffers from the approximation results which may be not valid for all circumstances so there are needs to modify the correlation between the rebound number and the concrete compressive strength under different conditions . Furthermore, the correlation curve can be created between the concrete compressive strength of standard cubes and the rebound number, nevertheless, the correlation can be different from testing to another according to ambient conditions , so that there is an urgent need to modify the correlation curve with different circumstances, such as different types of concrete or conditions can cause various characteristics, Kim also demonstrated that the increase in the surface hardness of the aggregate and carbonation degree also can effect on the Schmidt rebound number (Kim et al,2009)[8].

NDT is not required only for determining the strength of the structural elements with low resistance versus seismic load but also for quality control, curing time, and diagnosis of the defects in the existing concrete elements (Naderi, 2007)[9].

And also the strength information can be emphasized by happening failures in the construction elements that can be prevented by in-place test (Kosmatka S, et al., 2003)[10].

For measuring the concrete compressive strength, the Schmidt Hammer depends upon the rebound number which can be gotten by the spring load mass impact on the surface of concrete, the higher number of rebound numbers indicates the harder surface on another meaning the higher concrete compressive strength.

The objective of the research is to get the comparison between the results of the concrete cubic compressive strength and NDT, For the NDT: the Rebound Hammer used in the test was Type classic concrete Hammer with energy impact equal to 2.207 Nm-. Shown in Figure 1, Figure 2 shows the Compression Testing Machine used in the test, By using the linear regression analysis we can get the correlation between them, although the rebound number is not related to the concrete compressive strength directly, the regression analysis can be used to get the relationship between the dependent (Rebound number) and independent variables (concrete compressive strength), and SPSS software program is one of the popular programs can be used in the regression analysis.

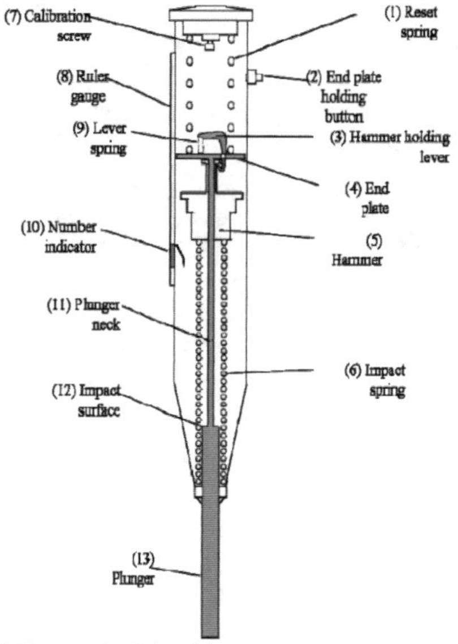

Figure 1. Schmidt Hammer

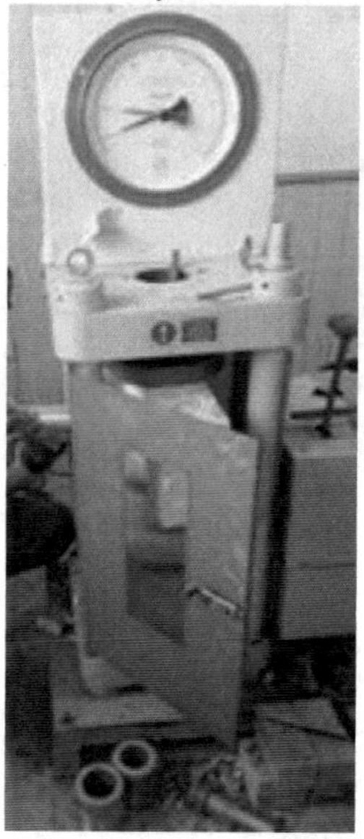

Figure 2. Compression machine test

The test of NDT has been done by the rebound of the elastic mass which depends on the hardness of the surface, in our testing the weight of the rebound hammer is 1.200 kg which is suitable to apply the test on the site, and the time of applied testing is at the end of curing after 28 days, the EN12504-02 determines the specification of the Schmidt test and number of reading for each specimen, furthermore, test of concrete cubic compressive strength can be carried out by the compression machine, the machine (Matest 1000 kN) was used in the experimental research, and a number of reading for each specimen is ten for different concrete grade for ordinary (200-299) Kg/cm^2, medium (300-399) Kg/cm^2, Simi high (400-499) Kg/cm^2 and high (over 50) Kg/cm^2 all tests have been carried after the duration of curing 28 days.

Different NDT can be used in place or on-site to determine not only the strength of concrete but also the cracks of the materials, voids behind the surface, the percentage of carbonation, or chloride et al (Breysse, 2012)[11].

The main standard for NDT methods is the surface hardness, ultrasonic pulse velocity, and penetration resistance methods (Malhotra et al., 2004). However, the purpose of the procedure of standard is to get the properties of the concrete on site.

The Schmidt Hammer test is considered as hardness test methods with indirect methods to determine the strength of concrete, there are different factors that can affect the values of rebound number such as the test area treatment, the component of the concrete element, type of cement, and aggregate, the position of the hammer, the humidity of the concrete, age of concrete and mix proportions (Jiri Brozovsky,2011)[12], And the results of the device is based on the impact of the hammer on the concrete by a spring, the results of testing depends on the hardness of concrete surface affected with a

Hammer and from these the concrete compressive strength is estimated via the surface rebound number.

Furthermore, the reliability of the nondestructive test has been investigated in various research (13),(14),(15)

All results of the research need to be careful in your recommendation which is based on the interpretation so, in our research, there are caring about the quality of instruments, the sample, and the statistical tool to get high accuracy and reliability.

3. Experimental

3.1. Materials

The characteristic of materials were used in the experimental to create concrete mixes as followings:

- *Aggregate*

The aggregate used in the experimental crushed basalt, crushed limestone and natural gravel (as a coarse aggregate), sand (as a fine aggregate) with properties as following in Tables 1 and Figure 3.

Table 1. Properties of Aggregate

Property	Sand	Gravel	Limestone	Basalt
Volume weight (t/m^3)	2.5	2.5	2.6	2.6
Specific gravity	1.6	1.5	1.6	1.7

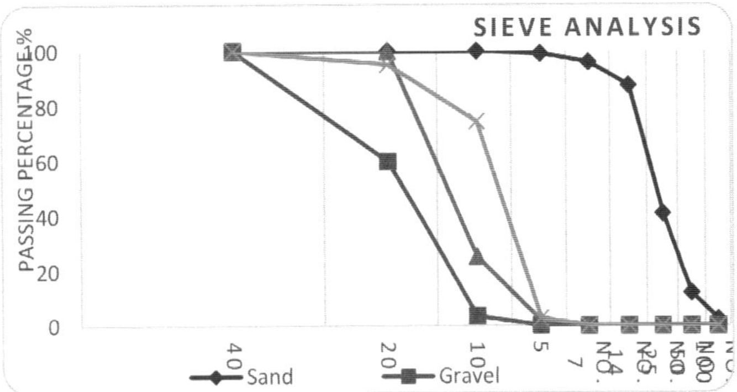

Figure 3. Aggregate Sieve Analysis

- *Cement*

Ordinary Portland Cement (OPC) was used as received. The chemical and physical properties of the cement are shown in Table 2

Table 2. Cement Properties.

Fineness of cement		3300 cm^2/g.	3300 cm^2/g
Strength of cement	2 day	248 Kg/cm^2	≥ 100Kg/cm^2
	28 day	425 Kg/cm^2	≥325 Kg/cm^2
Setting time	Initial	2:12 h	≥75 min
	Final	4:45 h	

- *Water*

The water used in the concrete mixtures is drinking water.

- *Concrete mixtures*

The test for studying the concrete strength after 28 days. Concrete mix trails proportion is shown in Table 3.

Table 3. Concrete mixes proportions by weight

Type of Agg.	Mix No.	Cement (Kg/m^3)	Coarse Agg.(Kg/m^3)	Sand (Kg/m^3)	Water (Liter/m^3)	Admixtures Liter/m^3)
Gravel	1	350	1266	590	175	0.50
Basalt	2	400	1334	614	137	0.34
Gravel	3	425	1198	539	213	0.50
Gravel	1	400	1194	556	200	0.50
Gravel	2	325	1697	799	166	0.51
Gravel	3	250	1411	657	125	0.50
Gravel	4	350	1379	919	165	0.47
Basalt	5	400	1334	621	132	0.33
Basalt	6	450	1273	593	136	0.30
Gravel	7	275	1374	640	135	0.49
Gravel	8	450	1120	524	225	0.50
Gravel	9	300	1702	783	167	0.56
Gravel	10	375	1229	572	187	0.50
Lime st.	11	400	1216	590	137	0.34
Basalt	12	400	1224	570	200	0.50
Basalt	13	400	1261	588	180	0.45
Lime st.	14	400	1216	590	133	0.33
Lime st.	15	450	1060	866	170	0.38
Lime st.	16	400	1216	590	135	0.34
Lime st.	17	450	1208	588	152	0.34
Basalt	18	500	1050	790	170	0.34
Basalt	19	400	1296	604	160	0.40
Lime st.	20	400	1160	540	200	0.50
Basalt	21	450	1273	593	152	0.34
Basalt	22	400	1334	621	140	0.35
Lime st.	23	400	1216	590	132	0.33
Basalt	24	450	1273	593	150	0.33
Lime st.	25	450	1208	588	150	0.33
Basalt	26	400	1334	614	135	0.34
Lime st.	27	450	1208	588	148	0.33
Gravel	28	350	1266	590	175	0.50
Basalt	29	400	1334	614	137	0.34
Gravel	30	425	1198	539	213	0.50

A super plasticizer was used (Sikament R2008, Sikament NN). Three standard cubes 15 cm x 15 cm x 15 cm were cast from each mix and cured in laboratory

3.2. Fabrication of tested samples

Using mixture of 0.1 m^3 capacity to compose the constituent materials .in the beginning, mix the dry aggregates, cement and aggregate for a one minute and then add the water which mixed with admixtures to create continuity mixing of the component until reaching to the homogenous constituent, this process will take two minutes to finish.

3.3. Instrumentation (Description of the testing machine)

Using the testing machine (MATEST to 100 tons up) to execute the static load which loaded from zero until failure.

4. Results and discussion

In the Experimental study , the cases of investigation were divided into four groups of concrete grades from ordinary, medium, Simi high to high grade , the total number of samples are 90s with different concrete mixtures design which sorting into four categories , the first group can be obtained with concrete compressive strength (200-299) Kg/cm^2, which had been done through 24 samples, the second groups can created with concrete compressive strength from (300-399) Kg/cm^2 with 18 samples , third group contains the samples with concrete compressive strength from (400-499) Kg/cm^2 with 24 samples and also the final group contains high grade of concrete compressive strength (over 500) Kg/cm^2 with 24 samples

Average of three cubes strength of each mix were calculated. Average of ten point of rebound Hammer test at each cube which subjected to 50% of Fcu failure were calculated.

So that all samples contain nearly different types of grade to create comprehensive comparison between results from rebound number Schmidt Hammer test NDT and Machine compression Test MCT, Tables (4,5,6,7) illustrate the values of the concrete compressive strength from manufacture curve of Rebound Hammer (RH) and average of three cubes values of the actual concrete compressive strength from MCT (Machine compression test).

Table 4. Compressive strength and rebound number for ordinary concrete grade (200-299) Kg/cm^2

Rebound Number	26.0	29.2	29.8	31.4	31.7	32.6	33.0	35.0
Fcu from Manufacturer's curve of R.H Kg/cm^2	175	225	230	245	250	280	280	305
Actual Fcu from MCT Kg/cm^2	175	223	237	241	248	271	275	298

Table 5. Compressive strength and rebound number for medium concrete grade (300-399) Kg/cm^2

Rebound Number	36.30	36.34	36.80	36.91	37.51	38.12
Fcu from Manufacturer's curve of R.H Kg/cm^2	315	320	340	345	360	370
Actual Fcu from MCT Kg/cm^2	324	326	339	345	352	363

Table 6. Compressive strength and rebound number for Simi high concrete grade (400-499) Kg/cm^2

Rebound Number	41.2	41.5	41.5	41.6	41.7	41.7	42.0	45.0
Fcu from Manufacturer's curve of R.H Kg/cm^2	410	425	425	427	430	430	440	490
Actual Fcu from MCT Kg/cm^2	406	420	420	427	433	433	444	493

Table 7. Compressive strength and rebound number for high concrete grade of (over 500) Kg/cm^2

Rebound Number	46.0	47.1	47.3	49.0	49.0	52.6	55.4	58.1
Fcu from Manufacturer's curve of R.H Kg/cm^2	500	520	525	560	560	620	680	780
Actual Fcu from MCT Kg/cm^2	515	528	530	568	568	623	736	813

Comparing the variations between the results of the concrete compressive strength for cubic compressive strength test MCT and non-destructive test (Schmidt Hammer) NDT.

All samples are cured for 28 days before getting observation of the results, main observation can be noticed that :

- Actual Fcu at until concrete grade 425 Kg/cm^2 give value lower than the R. H. T. Fcu but by increasing the concrete grade we found that the Actual Fcu give value higher than R. H. T. Fcu.
- The balance Fcu value which Actual Fcu equal R. H. T. Fcu was 427 kg/cm^2.
- So the using of Schmidt Hammer test need correlation equation for concrete grade until 425 kg/cm^2 and different correlation equation for concrete grade above of 427 kg/cm^2.

Tables 8 illustrates the comparison between the results of the average actual concrete compressive strength from Machine compression Test MCT and the average value from the Manufacturer's curve of R.H (Rebound Hammer test) ,Statistical T test had been done for analysis of the results , it contains two main hypothesis , the null and alternative hypothesis , the null hypothesis means there are not significant difference between the values of concrete compressive strength from RH (Rebound Hammer) and Machine compression Test MCT, and for the alternative hypothesis there are significant in the results between two investigation , the Paired T test is used to get the difference between the measurements from MCT and NDT for the same samples , the purpose of the statistical analysis is to determine the difference between paired experimental results , in the statistical analysis the acceptance of the null hypothesis depends on the values P to be bigger than 0.05 and if not the alternative hypothesis becomes acceptance , From table 8 it can obviously that :

- In case of ordinary grades : There aren't significant difference between the results from MCT and NDT
- In case of Medium grades : There aren't significant difference between the results from MCT and NDT
- In case of Semi High grades : There aren't significant difference between the results from MCT and NDT
- In case of High grades : There are significant difference between the results from MCT and NDT

Table 8. Statistical T test analysis and results.

Sample Grade	P-value vs. Level of Significance, α	conclusion
Ordinary grade (200-299) Kg/cm^2	P-value = 0.156 α = 0.05 P-value > α; The null hypothesis is accepted and Fail to replace by alternative hypothesis	There aren't significant difference between the results from MCT and NDT and the average strength generated from the Manufacturer's curve of R.H underestimates by 1.12% the average actual concrete compressive strength from MCT
Medium grade (300-399) Kg/cm^2	P-value = 0.954 α = 0.05 P-value > α; The null hypothesis is accepted and fail to replace by alternative hypothesis	There aren't significant difference between the results from MCT and NDT and the average strength generated from the Manufacturer's curve of R.H underestimates by 0.05% the average actual concrete compressive strength from MCT
Simi high grade (400-499) Kg/cm^2	P-value = 0.931 α = 0.05 P-value > α; The null hypothesis is accepted and fail to replace by alternative hypothesis	There aren't significant difference between the results from MCT and NDT and there aren't nearly difference between average strength generated from the Manufacturer's curve of R.H and the average actual concrete compressive strength from MCT
High grade of (over500) Kg/cm^2	P-value = 0.035 α = 0.05 P-value < α; The null hypothesis is rejected, and the hypothesis will be in favor of alternative hypothesis	There are significant difference between the results from MCT and NDT and the average strength generated actual concrete compressive strength from MCT by 2.87 % the Manufacturer's curve of R.H

4.1. *The relationship between Schmidt Hammer Rebound Number and Actual concrete Compressive Strength***:**

Linear regression analysis can be used to get the correlation curve between the variable values of the rebound number and the actual concrete compressive strength, the correlation coefficient R2 for testing samples, the regression analysis is a series of statistical processes for creation the relationship, for the study analysis the correlation curves for the different cases of grades can be illustrated by Figures (4&5&6&7).

In Figure 4 (Ordinary Grade Case) : it is observed that the correlation curve is very close to the variables data of RN and Fcu from MCT , the high value of R square ($R^2 = 0.976$) is the prove for the correlation

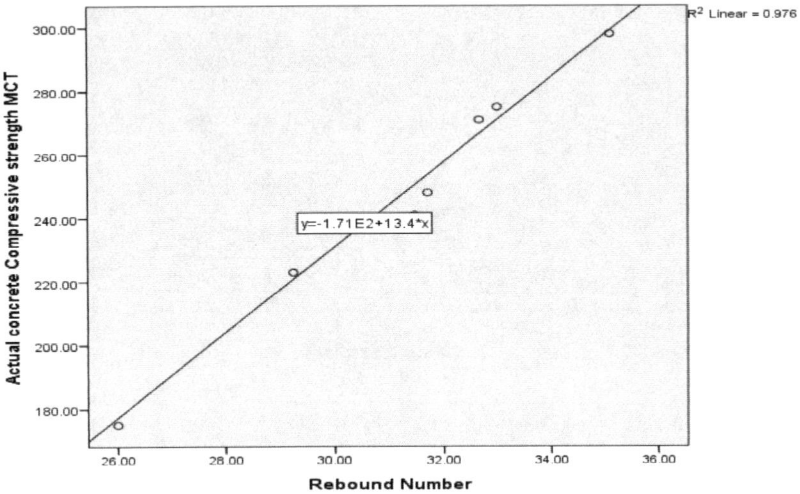

Figure 4. Actual Concrete Compressive strength From MCT vs. Rebound hammer in case of Low Grade

In Figure 5 (Medium Grade Case) : it is observed that the correlation curve is very close to the variables data of RN and Fcu from MCT , the high value of R square ($R^2 = 0.970$) is the prove for the correlation

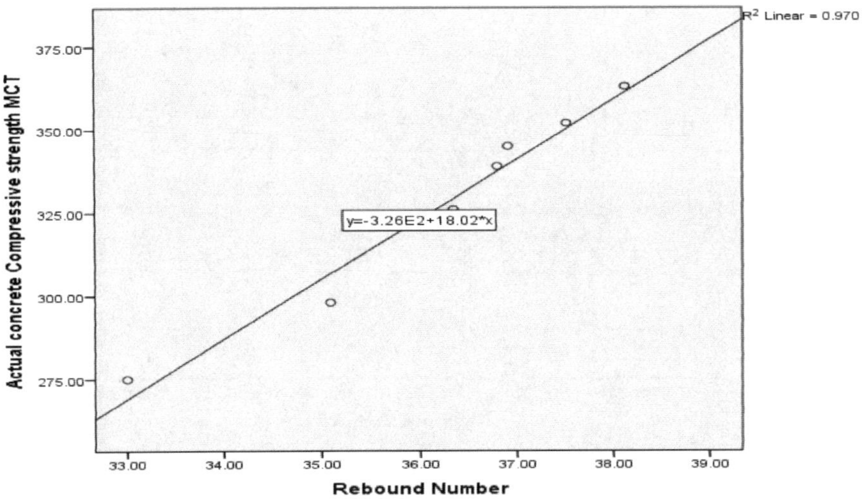

Figure 5. Actual Concrete Compressive strength From MCT vs. Rebound hammer in case of Medium Grade

In Figure 6 (Semi High Grade Case) : it is observed that the correlation curve is very close to the variables data of RN and Fcu from MCT , the high value of R square ($R^2 = 0.933$) is the prove for the correlation

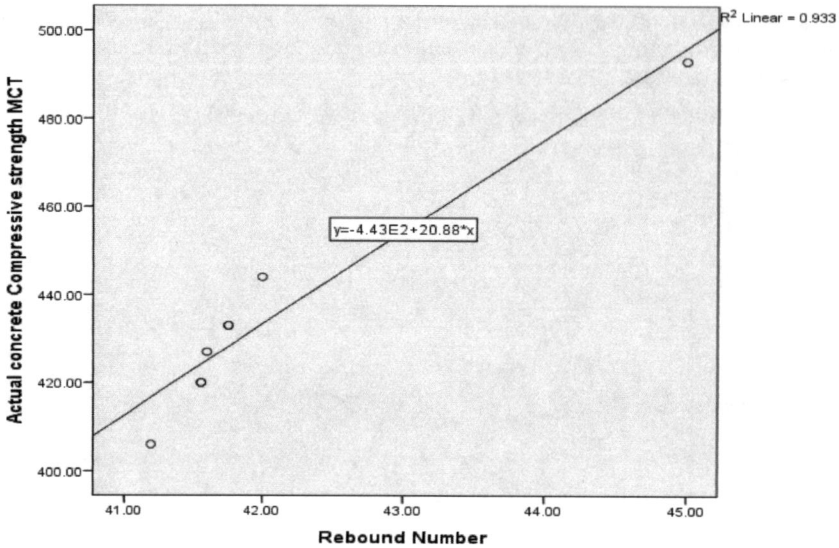

Figure 6. Actual Concrete Compressive strength From MCT vs. Rebound hammer in case of Semi High Grade

In the final Figure 7 (High Grade Case) : it is observed that the correlation curve is very close to the variables data of RN and Fcu from MCT , the high value of R square ($R^2 = 0.974$) is the prove for the correlation

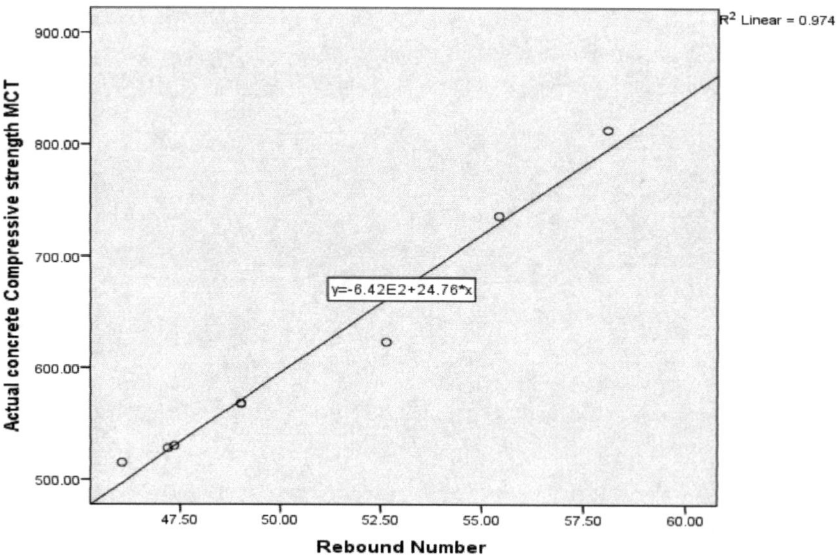

Figure 7. Actual Concrete Compressive strength From MCT vs. Rebound hammer in case of Very High Grade

4.2. The comparison in the relationship between values of concrete compressive strength from Schmidt Hammer and the Machine compression test

Figure 8 illustrates the relationship between the values of concrete compressive strength which gotten from the Schmidt Hammer and the machine compression test.

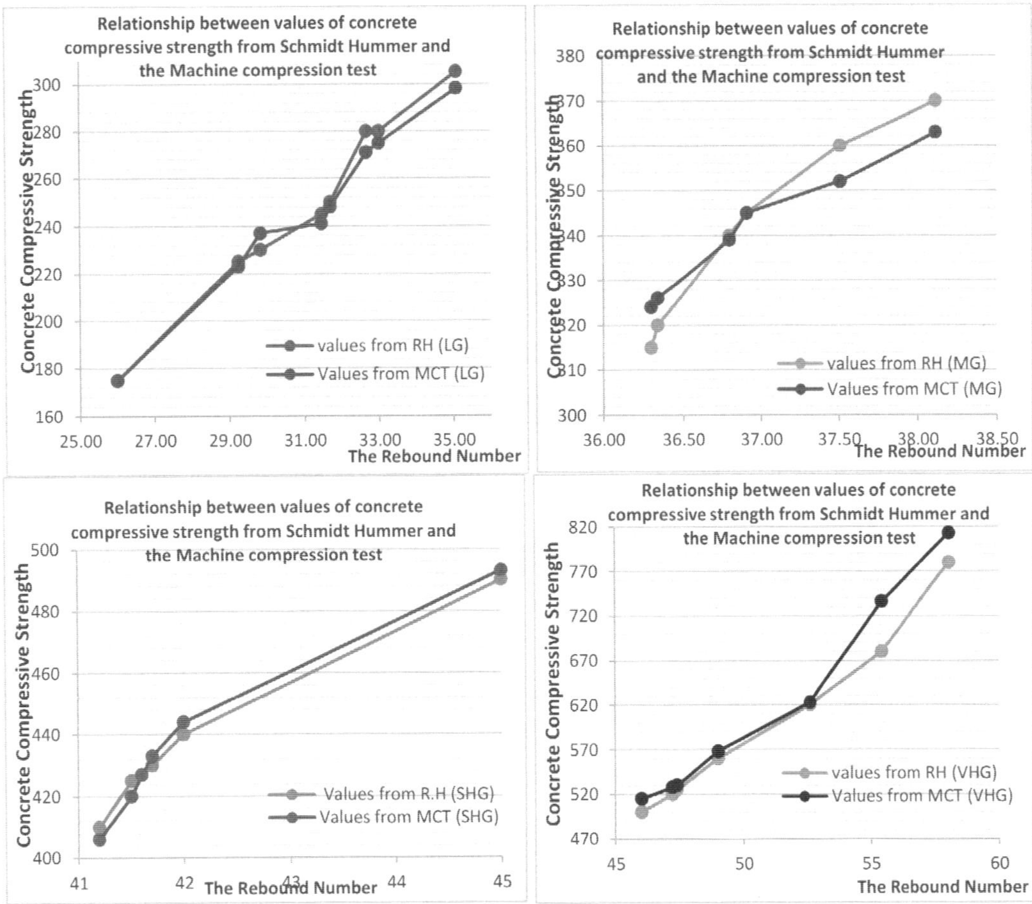

Figure 8. Relationship between values of concrete compressive strength from Schmidt Hammer and the Machine compression test

5. Conclusion

In the final conclusion, using the statistical analysis to get the full relationship between the manufacturing curve of NDT and the actual concrete compressive strength from the machine compression test, using Paired T-test can help to get the variances between values from RH and MCT, and from the analysis, the variances between them in case of Ordinary. the medium and semi-high grade is very low which means the null hypothesis is accepted according to the statistical analysis and on another side, there are variances in case of high grade according to statistical analysis,

The study results notice that the relationship between values of Cubic compressive strength from Rebound Hammer No. and the Machine compression test can achieved by using relation formula as shown in Figure (4,5,6&7) for each grade.

The study also interested to get from results; it is obviously the closing in the values in high accuracy level with extreme confidence.

References

[1] Yılmaz, N., & Avşar, Ö. (2013). Structural damages of the May 19, 2011, Kütahya–Simav earthquake in Turkey. Natural hazards, **69**(1), 981-1001.

[2] Abrams, D. A. (1927, February). Water-cement ratio as a basis of concrete quality. In Journal Proceedings (Vol. 23, No. 2, pp. 452-457).

[3] Yon, B., Sayin, E., Calayir, Y., Ulucan, Z. C., Karatas, M., Sahin, H., ... & Bildik, A. T. (2015). Lessons learned from recent destructive Van, Turkey earthquakes. Earthquakes and Structures, **9**(2), 431-453.

[4] Alyamac, K. E., Ghafari, E., & Ince, R. (2017). Development of eco-efficient self-compacting concrete with waste marble powder using the response surface method. Journal of Cleaner Production, **144**, 192-202.

[5] Malhotra, V. M., & Carino, N. J. (2004). Handbook on nondestructive testing of concrete. CRC press.

[6] İlhan, İ. What does the Schmidt Hammer do?, TMH-Turkisn Engineering News. Issue **410**, pp. 26-28, 2000

[7] Hobbs, B., & Kebir, M. T. (2007). Non-destructive testing techniques for the forensic engineering investigation of reinforced concrete buildings. Forensic science international, **167**(2-3), 167-172.

[8] Kim, J. K., Kim, C. Y., Yi, S. T., & Lee, Y. (2009). Effect of carbonation on the rebound number and compressive strength of concrete. Cement and Concrete Composites, **31**(2), 139-144.

[9] Naderi, M. (2007). New twist-off method for the evaluation of in-situ strength of concrete. Journal of Testing and Evaluation, **35**(6), 602-608.

[10] Kosmatka, S. H., Kerkhoff, B., & Panarese, W. C. (2003). Design and control of concrete mixtures (Vol. **5420**, pp. 60077-1083). Skokie, IL: Portland Cement Association.

[11] Breysse, D. (2012). Nondestructive evaluation of concrete strength: An historical review and a new perspective by combining NDT methods. Construction and Building Materials, **33**, 139-163.

[12] Brozovsky, J., & Zach, J. (2011, October). Influence of surface preparation method on the concrete rebound number obtained from impact hammer test. In Pan American Conference for NDT (Vol. **6**, No. 10).

[13] Abdul Rahim M, Shahidan S, Onn LC, Saiful Bahari NAA, Abd Rahman N, Ayob A. (2020). The Behavior of Non-Destructive Test for Different Grade of Concrete. IJIE International Journal of Integrated Engineering (12)

[14] Olonade KA. (2020). influence of water-cement ratio and water reducing admixtures on the rebound number of hardened concrete. MJCE Malaysian Journal of Civil Engineering (32)

[15] Kumavat HR, Chandak NR. (2021). Repeatability of concrete rebound hardness using comprehensive statistical analysis. JKSUES Journal of King Saud University - Engineering Sciences.

Defected Photonic Crystal Array Using Porous GaN as Malaria Sensor

M.T. Tammam[1], Zaky A. Zaky*[,1] ,Arvind Sharma[2], Z.S.Matar[3], Arafa H.Aly[1,*,] M.A. Mohaseb[1,3]
[1]TH-PPM Group, Physics Department, Faculty of Sciences, Beni-Suef University, Beni-Suef, Egypt
[2]Department of Physics, Government Dungar College, Bikaner, Rajasthan, India -334001
[3]Umm-Al-Qura University, Faculty of Applied Science, Department of Physics, Mecca, Saud Arabia
*Corresponding authors. E-mail: zaky.a.zaky@science.bsu.edu.eg (Zaky A. Zaky)
arafaaly@aucegypt.edu (Arafa H. Aly)

Abstract. A defective one-dimensional photonic crystal is investigated as a biosensor to detect malaria disease. The proposed photonic structure is air/(GaN/Porous GaN)^N/Sample/(GaN/Porous GaN)^N/Substrate. The red blood cells sample of the human being is used as a sample defect in the proposed optical device. The pioneer transfer matrix method is used to analyze the transmittance spectra. A change in sample refractive index highly affects the transmittance resonant peak and this shift in the peak plays a key role in the operation of the device. The relatively high figure of merit of 1022 RIU^{-1} with a sensitivity of 1472 nm/RIU and quality factor of 1076 is detected. The proposed sensor is relatively better than others available to detect malaria disease.

Keywords:
Photonic Crystal, Biosensor, Porous materials, Malaria Diagnosis

1. Introduction

In the last few decades, defected one dimensional photonic crystal (1D-PC) have attracted more attention in many potential applications such as biosensors [1-11], temperature sensors [12-16],solar cells [17-22], filters [23-30], air purifier [31], water desalination [32, 33] and others [34-36]. The most amazing feature of 1D-PC is the photonic bandgaps (PBG) originated due to multiple scattering at interfaces, and the incident waves are blocked to propagate within the PBGs [37, 38]. Because of the resonance of Fabry-Perot, a defect resonant peak appears within the PBG when a defect layer is introduced into 1D-PC [39]. As a result of the law of Bragg Snell's, by changing the index of refraction of the defect layer, the resonant peak will be shifted. This dependence between the optical properties of the sample layer sample and the resonant peak is the principle of PC sensors.

Recently, Porous and two-dimensional layers have gained the attention of scholars in biosensing applications [40-42]. Defected 1D-PC of porous layers can sense different solutions based on their refractive indices change. Porous Gallium nitride (GaN) can be used as a multilayer PC with perfect lattice matching [43-45]. Porous GaN is more interesting than porous dielectrics because it can easily be incorporated with active electronics [42]. The optical properties of GaN can be adjusted by changing the porosity [46]. Lheureux et al. [43] used standard photolithography to prepare the porous GaN utilizing the wet EC etching procedure.

Malaria is a serious disease that might kill 405,000 persons in 2018 according to WHO report [47]. Early diagnosis contributes to the high speed of achieving a cure for the disease. Malaria can be diagnosed using a variety of approaches, including quantitative buffy coat (QBC) method [48], microscopic diagnosis [49], rapid diagnostic tests (RDT) [50], indirect fluorescent antibody (IFA)[51], and others [52]. Most of these techniques have limitations such as not detect early stages effectively [48], take a large testing time [53], and not being cost-effective [54]. Human tissues are injected with a parasite with a mosquito bite during its sporozoite phase and cause malarial infection. This parasite moves through various stages (ring, trophozoite, and schizont) in red blood cells (RBCs). These phrases cause a change in the indices of refraction of the infected RBCs, as clear in Table 1. So, detecting the index of refraction of the RBCs is considered the key factor for malaria detection [55, 56].

Table 1: Average index of refraction values of different cells of RBCs [57]:

Name	Average refractive index
Normal RBC cells	1.402
Ring phase cells	1.395
Trophozoite phase cells	1.383
Schizont phase cells	1.373

In this paper, we will try to overcome such limitations with defected 1D-PC malaria sensor using Porous GaN. The proposed structure has the potential to aid in the early detection of malaria.

2. Materials and method

The proposed structure of the malaria sensor is air/(GaN/Porous GaN)^N/Sample/(GaN/Porous GaN)^N/Substrate as clear in Fig.1. At the center of the structure, a defect layer is inserted between two identical 1D-PC of (GaN/Porous GaN)^N. N is the total number of layers of 1D-PC. A transverse electric (TE) polarization will fall on the structure. The thicknesses of the first and second layers are d_1 =163 nm and d_2=208 nm, with refractive indices as in Eq. 1 and Eq. 2. This porosity of porous GaN will be 53 % [43]. Experimentally and theoretically, a multilayer of GaN/Porous GaN was studied [43, 45, 58]. As clear in Fig. 1, by injecting the sample at the top of the structure, the pores will be infiltrated with the analyte sample. The index of refraction of the porous GaN (n_2) is calculated by using volume average theory [46]:

$$n_2 = \sqrt{(1-P)n_1^2 + Pn_{sam}^2}, \quad (1)$$

where P, n_{sam}, n_1, are the porosity ratio, the index of refraction of the analyte that is infiltrated inside the pores (index of the RBCs sample), and the index of refraction of GaN. The index of refraction of GaN as a function of wavelength (λ) is calculated as [59]:

$$n_1 = \sqrt{3.6 + \frac{1.75\lambda^2}{\lambda^2 - 0.256^2} + \frac{4.1\lambda^2}{\lambda^2 - 17.86^2}}. \quad (2)$$

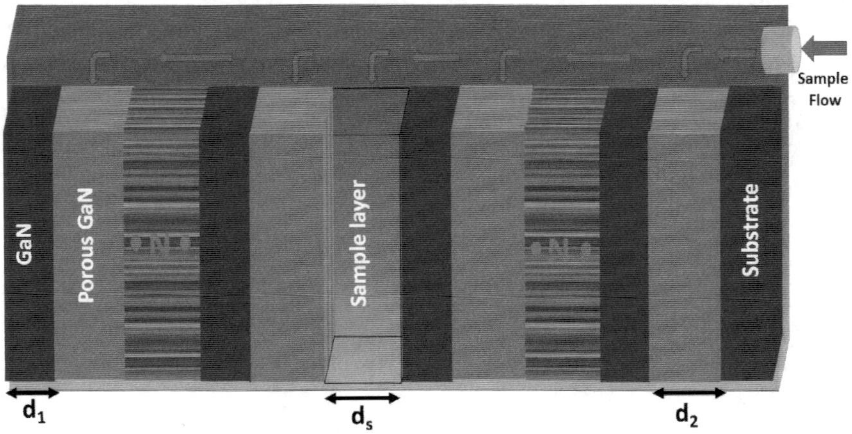

Figure 1. The structure of the proposed 1D-PC sensor.

Multilayer structure problems can be solved and analyzed by the famous transfer matrix method (TMM). The TMM is explained by several authors [60-63]. The transmittance of the proposed device is calculated using TMM. Here we are considering the interaction between incident transverse electromagnetic (TE) waves with normal incidence. It is given by the overall matrix as follows-

$$C = \begin{vmatrix} C_{11} & C_{12} \\ C_{21} & C_{22} \end{vmatrix} = (c_1 c_2)^N c_{sam} (c_1 c_2)^N , \quad (3)$$

where C_{11}, C_{12}, C_{21} and C_{22} are elements of the transfer matrix. Here c_1, c_2 and c_{sam} are characteristic matrix corresponding to GaN, porous GaN, and sample which are as follows:

$$c_k = \begin{pmatrix} cos\emptyset_k & -\frac{i}{p_E} sin\emptyset_k \\ -ip_E sin\emptyset_k & cos\emptyset_k \end{pmatrix}, \quad (4)$$

where k= 1, 2, and sam. The $\emptyset_k$ is phase difference at each layer and it is denoted by:

$$\emptyset_k = \frac{2\pi n_k d_z cos\emptyset_k}{\lambda}. \quad (5)$$

The values of p_k for the TE(S) polarisation is given by $p_k = n_k cos(\theta_k)$.

The incident angle $\theta_0, \theta_1, \theta_2$, θ_{sam}, and $\theta_s$are at the surfaces of air, GaN, porous GaN, sample, and substrate. Here angle θ_0 represents the angle of incidence of the incident light from the air to the proposed device and satisfying the Snells law as follows:

$$n_0 sin(\theta_0) = n_1 sin(\theta_1) = n_2 sin(\theta_2) = n_{sam} sin(\theta_{sam}) = n_s sin(\theta_s). \quad (6)$$

The matrix $(c_1 c_2)^N$ can be calculated by using the Chebyshev polynomials of the second kind. The transmission coefficient (t) using the above equations is given below[48-50] :

$$t = \frac{2p_s}{(C_{11} + C_{12}p_0)p_s - (C_{21} + C_{22}p_0)}. \quad (7)$$

Hence, the transmittance (T) of the proposed device is given by,

$$T(\%) = 100 * \frac{p_0}{p_s} [t]^2. \quad (8)$$

3. Results and discussions

The proposed structure of the malaria sensor is as air/(GaN/Porous GaN)^N/Sample/(GaN/Porous GaN)^N/Substrate. At the middle of the structure, a defect red blood cells sample is inserted between two identical 1D-PC of (GaN/Porous GaN)^N. A transverse electric (TE) polarization will be incident on the structure. The transmittance of the proposed sensor without defect layer (black line), and with defect layer (red line) at n_{sam}=1.731 is depicted in Fig.2. Without defect layer resonant peak non exists but using the defect layer of refractive index 1.731 a clear sharp resonant peak of 100% and large photonic bandgap exist.

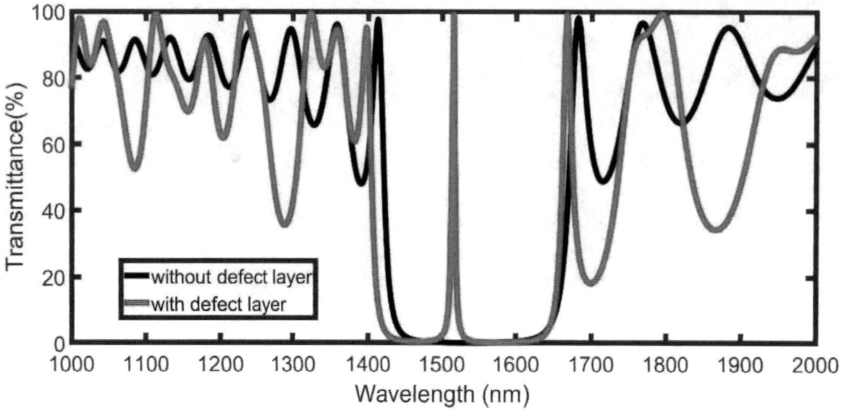

Figure 2. The transmittance of the proposed sensor without defect layer (black line), and with defect layer (red line) at n_{sam}=1.731.

The sensitivity (S) of any sensor, figure of merit (FoM), and Q factor can be calculated as [64-66]:

$$S = \frac{\Delta\lambda_R}{\Delta n_{sam}}, \quad (9)$$

$$FoM = \frac{S}{FWHM}, \quad (10)$$

$$Q = \frac{\lambda_R}{FWHM} \quad (11)$$

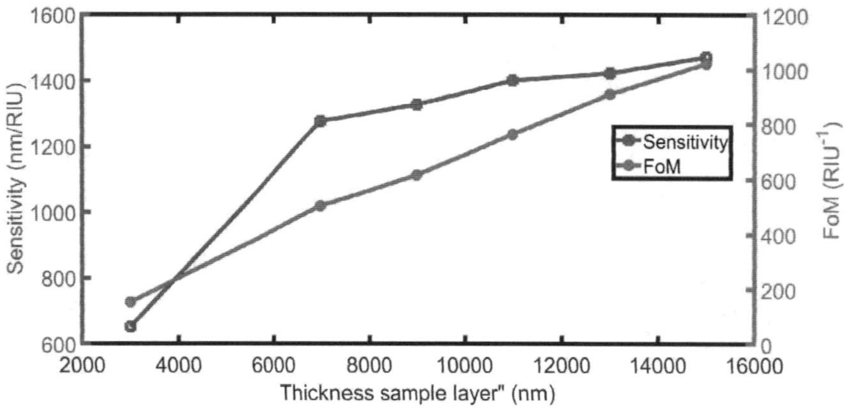

Fig. 3: The sensitivity and FoM at different thicknesses of the sample layer.

The performance parameters of the investigated sensor at different thicknesses of the sample layer are clear in Fig.3. The rise in the defect layer of the sample analyte thickness raises the sensitivity of the photonic sensor. The rise in defect of the sample analyte layer thickness from 3000 nm to 15000 nm raises sensitivity value from 652.3 nm/RIU to 1472.2 nm/RIU. Initially, the rise in sensitivity value is sharper. If defect layer thickness increases onwards 11000 nm, the sensitivity is approximately constant. The variation in the position of resonance mode is that it depends upon the total index of refraction of all structures and thickness of periodic PC. The standing wave formation due to defect at the resonant wavelength obey the condition [4, 67]

$$O = z\lambda = n_{eff}g, \qquad (12)$$

where O is the optical path length, z is a scalar bomber, λ is the incident wavelength, n_{eff} is the refractive index of the whole device and g is the path length. Here, we are increasing g but z and n_{eff} remain constant hence the resonant wavelength shift towards a longer wavelength. The FoM increases approximately linearly from 152.3 RIU^{-1} to 1022.6 RIU^{-1} with a rise from 3000 nm to 15000 nm in the sample layer thickness, respectively. It is clear from the investigation that the optimum value of sample layer thickness obtained was 15000 nm for different RBCs. The human tissues are injected with a parasite with a mosquito bite during its sporozoite phase and cause malarial infection. This parasite goes through different phases in RBCs. These phrases cause a change in the indices of refraction of the infected RBCs, as clear in Table 1. The transmittance of the proposed sensor with an optimum sample layer thickness of 15000 nm for different phases in RBCs is depicted in Fig.4. For the phases normal RBC, ring, trophozoite, and schizont in red blood cells (RBCs) the refractive index is decreasing. For the phases ring, trophozoite, and schizont in red blood cells (RBCs) concerning normal RBC, the resonant peak shifted towards lower wavelength (blue shift) due to decreasing values of refractive index.

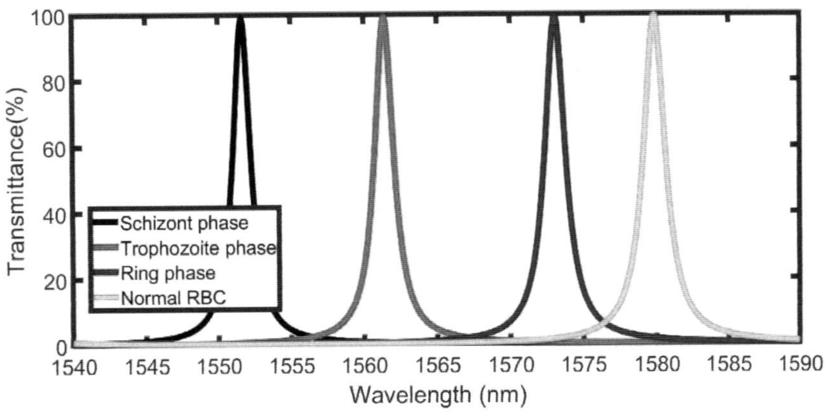

Figure 4. The transmittance of the investigated sensor with a sample layer thickness of 15000 nm for different RBCs.

The performance comparison of our work is represented in the following table 2. The relatively high sensitivity of 1472 nm/RIU with a figure of merit of 1022 RIU^{-1} and quality factor 1076 is detected.

Table 2: Performance comparison (NC= not be calculated).

Reference	S (nm / RIU)	FoM (RIU^{-1})	Q-factor
2017, [68]	17	233	$3*10^4$
2019, [6]	1000	----	35 517
2020, [2]	1400	$6*10^6$	$3.5*10^6$
2020, [69]	777	----	2576
2020, [70]	656	1.13	2719
2020, [43]	10	15.1	300
2021, [52]	496	----	203000
This work	**1472**	**1022**	**1076**

4. Conclusion

A GaN and porous GaN-based one-dimensional defective photonic crystal is investigated as an optical sensor to detect malaria disease. The red blood cells sample of the human being is used as a sample defect in the proposed optical sensor. The notable change in sample refractive index affects the transmission resonant peak and the shift in the peak plays a vital role in the operation of the photonic sensor. The relatively high sensitivity of 1472 nm/RIU with a figure of merit 1022 RIU^{-1} and quality factor 1076 was detected. The proposed optical sensor is relatively better than others available to detect malaria disease.

Conflicts of interest/Competing interests

The authors declare no conflicts of interest.

ORCID IDs

https://orcid.org/0000-0003-0795-378X (A. H. Aly)
https://orcid.org/0000-0002-8300-7755 (Z. A. Zaky)

References

[1] Zaky Z A and Aly A H 2021 Modeling of a biosensor using Tamm resonance excited by graphene *Appl. Optics* **60** 1411-9

[2] Aly A H, Zaky Z A, Shalaby A S, Ahmed A M and Vigneswaran D 2020 Theoretical study of hybrid multifunctional one-dimensional photonic crystal as a flexible blood sugar sensor *Phys. Scr.* **95** 035510

[3] Ramanujam N R, El-Khozondar H J, Dhasarathan V, Taya S A and Aly A H 2019 Design of one dimensional defect based photonic crystal by composited superconducting material for bio sensing applications *Physica B* **572** 42-55

[4] Aly A H and Zaky Z A 2019 Ultra-sensitive photonic crystal cancer cells sensor with a high-quality factor *Cryogenics* **104** 102991

[5] Aly A H, Mohamed D, Zaky Z A, Matar Z S, Abd El-Gawaad N S, Shalaby A S, Tayeboun F and Mohaseb M 2021 Novel Biosensor Detection of Tuberculosis Based on Photonic Band Gap Materials *Mater. Res.-Ibero-am. J. Mater.* **24** e20200483

[6] Arunkumar R, Suaganya T and Robinson S 2019 Design and analysis of 2D photonic crystal based biosensor to detect different blood components *Photonic Sensors* **9** 69-77

[7] Abd El-Ghany S E, Noum W M, Matar Z, Zaky Z A and Aly A H 2020 Optimized bio-photonic sensor using 1D-photonic crystals as a blood hemoglobin sensor *Phys. Scr.* **96** 035501

[8] Banerjee A 2020 Design of enhanced sensitivity gas sensors by using 1D defect ternary photonic band gap structures *Indian J. Phys.* **94** 535-9

[9] Qutb, S R,Aly, A H, Sabra, W, 2021 Salinity and temperature detection for seawater based on a 1D-defective photonic crystal material International Journal of Modern Physics B, Vol. 35, No. 1 (**2021**) 2150012-13

[10] Ahmed A M and Shaban M 2020 Highly sensitive Au–Fe_2O_3–Au and Fe_2O_3–Au–Fe_2O_3 biosensors utilizing strong surface plasmon resonance *Appl. Phys. B-Lasers Opt.* **126** 1-10

[11] Aly A H, Mohamed D, Mohaseb M, El-Gawaad N S A and Trabelsi Y 2020 Biophotonic sensor for the detection of creatinine concentration in blood serum based on 1D photonic crystal *RSC Adv.* **10** 31765-72

[12] Amiri I S, Paul B K, Ahmed K, Aly A H, Zakaria R, Yupapin P and Vigneswaran D 2019 Tri-core photonic crystal fiber-based refractive index dual sensor for salinity and temperature detection *Microw. Opt. Technol. Lett.* **61** 847-52

[13] Zaky, Z A, Aly, A H 2021 Highly Sensitive Salinity and Temperature Sensor Using Tamm Resonance, *Plasmonics*, https://doi.org/10.1007/s11468-021-01487-6

[14] Banerjee A 2009 Enhanced temperature sensing by using one-dimensional ternary photonic band gap structures *Progress In Electromagnetics Research* **11** 129-37

[15] Zaky Z A, Ahmed A M and Aly A H 2021 Remote Temperature Sensor Based on Tamm Resonance *Silicon* https://doi.org/10.1007/s12633-021-01064-w

[16] Aly, A H, Abdel Ghany, S E-S,, Fadlallah, M M, Salman, F E, and Kamal, B M 2015 Transmission and Temperature Sensing Characteristics of a Binary and Ternary Photonic Band Gap, J. Nanoelectron. Optoelectron. 10, 9-14.

[17] Aly A H and Sayed H 2017 Enhancement of the solar cell based on nanophotonic crystals *J. Nanophotonics* **11** 046020

[18] Zhang X-L, Song J-F, Li X-B, Feng J and Sun H-B 2012 Optical Tamm states enhanced broad-band absorption of organic solar cells *Appl. Phys. Lett.* **101** 243901

[19] Aly A H and Sayed H 2018 Photonic band gap materials and monolayer solar cell *Surf. Rev. Lett.* **25** 1850103

[20] De la Mora M, Jaramillo O, Nava R, Tagüeña-Martínez J and Del Río J 2009 Viability study of porous silicon photonic mirrors as secondary reflectors for solar concentration systems *Solar energy materials and solar cells* **93** 1218-24

[21] Aly A H, Sayed H and Elsayed H A 2019 Development of the Monolayer Silicon Solar Cell Based on Photonic Crystals *Silicon* **11** 1377-82

[22] Sayed H and Aly A H 2021 Salinity optical sensor by using two-dimensional photonic crystals: computational study *Mat. Sci. Eng. B.* **269** 115169

[23] Aly A H, Ahmed A M and Shaban M 2020 Multilayer angular optical filter as a smart window *Indian J. Phys.* **94** 95-103

[24] Pandey A, Tripathi A, Srivastava S and Jit S 2017 Analysis of ternary layer photonic band gap tunable filters for wavelength division multiplexing applications *J. Nanoelectron. Optoelectron.* **12** 331-6

[25] Sabra W, Azzam S I, Song M, Povolotskyi M, Aly A H and Kildishev A V 2018 Plasmonic metasurfaces for subtractive color filtering: optimized nonlinear regression models *Opt. Lett.* **43** 4815-8

[26] Awasthi S K and Ojha S P 2008 Design of a tunable optical filter by using a one-dimensional ternary photonic band gap material *Progress in Electromagnetics Research* **4** 117-32

[27] Aly A H, Barakat S and Amin A 2020 Tunable filter based on the 1D photonic crystal within ultraviolet radiations *IOP Conf. Ser.: Mater. Sci. Eng.* **956** 012010.

[28] Usievich B, Prokhorov A and Sychugov V 2002 A photonic-crystal narrow-band optical filter *Laser Phys.* **12** 898-902

[29] Aly A H, Ameen A A, Elsayed H A, Mohamed S and Singh M R 2019 One-dimensional metallo-superconductor photonic crystals as a smart window *J. Supercond. Nov. Magn* **32** 2313-8

[30] Aly A H, Ameen A A and Vigneswaran D 2019 Superconductor nanometallic photonic crystals as a novel smart window for low-temperature applications *J. Supercond. Nov. Magn* **32** 191-7

[31] Ameen A A, Elsayed H and Aly A H 2021 Towards a highly efficient air purifier using annular photonic crystals in UV regimes *RSC Adv.* **11** 14915-21

[32] Sayed H, Krauss T F and Aly A H 2020 Versatile photonic band gap materials for water desalination *Optik* **219** 165160

[33] Trabelsi Y, Belhadj W, Ben Ali N, Aly A H 2021 Theoretical Study of Tunable Optical Resonators in Periodic and Quasiperiodic One-Dimensional Photonic Structures Incorporating a Nematic Liquid Crystal. *Photonics*, *8*, 150.

[34] Aly A H, Mohamed D, Matar Z, Trabelsi Y, Vigneswaran D, Tayeboun F and Mohaseb M 2021 Tunability and Fano Resonance Properties in Different Types of One-Dimensional Superconductor Photonic Crystals, *Mat. Res.* **24(4)**

[35] Malek C, Aly A H and Sabra W 2021 Tunable PBGs with a cutoff frequency feature in Fibonacci quasi-periodic designs containing a superconductor material at THz region *Phys. Scr. 96, 105501.*

[36] Aly A H, Shaban S M and Mehaney A 2021 High-performance phoxonic cavity designs for enhanced acousto-optical interaction *Appl. Optics* **60** 3224-31

[37] Yablonovitch E 2001 Photonic crystals: semiconductors of light *Sci.Am.* **285** 46-55

[38] Yablonovitch E and Gmitter T 1989 Photonic band structure: The face-centered-cubic case *Phys. Rev. Lett.* **63** 1950

[39] van Popta A C, Hawkeye M M, Sit J C and Brett M J 2004 Gradient-index narrow-bandpass filter fabricated with glancing-angle deposition *Opt. Lett.* **29** 2545-7

[40] Agarwal V, Del Río J, Malpuech G, Zamfirescu M, Kavokin A, Coquillat D, Scalbert D, Vladimirova M and Gil B 2004 Photon Bloch oscillations in porous silicon optical superlattices *Phys. Rev. Lett.* **92** 097401

[41] Hidalgo N, Calvo M E, Colodrero S and Miguez H 2010 Porous one-dimensional photonic crystal coatings for gas detection *IEEE Sens. J.* **10** 1206-12

[42] Wang Z, Zhang J, Xu S, Wang L, Cao Z, Zhan P and Wang Z 2007 1D partially oxidized porous silicon photonic crystal reflector for mid-infrared application *J. Phys. D-Appl. Phys.* **40** 4482

[43] Lheureux G, Monavarian M, Anderson R, DeCrescent R A, Bellessa J, Symonds C, Schuller J A, Speck J, Nakamura S and DenBaars S P 2020 Tamm plasmons in metal/nanoporous GaN distributed Bragg reflector cavities for active and passive optoelectronics *Opt. Express* **28** 17934-43

[44] Yerino C D, Zhang Y, Leung B, Lee M L, Hsu T-C, Wang C-K, Peng W-C and Han J 2011 Shape transformation of nanoporous GaN by annealing: From buried cavities to nanomembranes *Appl. Phys. Lett.* **98** 251910

[45] Zhu T, Liu Y, Ding T, Fu W Y, Jarman J, Ren C X, Kumar R V and Oliver R A 2017 Wafer-scale fabrication of non-polar mesoporous GaN distributed Bragg reflectors via electrochemical porosification *Sci. Rep.* **7** 1-8

[46] Zhang C, Park S H, Chen D, Lin D-W, Xiong W, Kuo H-C, Lin C-F, Cao H and Han J 2015 Mesoporous GaN for Photonic Engineering Highly Reflective GaN Mirrors as an Example *ACS photonics* **2** 980-6

[47] https://www.who.int/malaria/ W r a a and publications/world-malaria-report-2019/en/

[48] Clendennen T E, Long G W and Baird J K 1995 QBC® and Giemsa-stained thick blood films: diagnostic performance of laboratory technologists *Transactions of the Royal Society of Tropical Medicine and Hygiene* **89** 183-4

[49] Erdman L K and Kain K C 2008 Molecular diagnostic and surveillance tools for global malaria control *Travel medicine and infectious disease* **6** 82-99

[50] Bell D, Wongsrichanalai C and Barnwell J W 2006 Ensuring quality and access for malaria diagnosis: how can it be achieved? *Nature Reviews Microbiology* **4** 682-95

[51] She R C, Rawlins M L, Mohl R, Perkins S L, Hill H R and Litwin C M 2007 Comparison of immunofluorescence antibody testing and two enzyme immunoassays in the serologic diagnosis of malaria *Journal of travel medicine* **14** 105-11

[52] Suthar B and Bhargava A 2021 Biosensor Application of 1D Photonic Crystal for Malaria Diagnosis *Plasmonics* **16** 59-63

[53] Sulzer A, Wilson M and Hall E C 1969 Indirect fluorescent-antibody tests for parasitic diseases *The American journal of tropical medicine and hygiene* **18** 199-205

[54] Hänscheid T and Grobusch M P 2002 How useful is PCR in the diagnosis of malaria? *Trends in parasitology* **18** 395-8

[55] Liu P Y, Chin L, Ser W, Chen H, Hsieh C-M, Lee C-H, Sung K-B, Ayi T, Yap P and Liedberg B 2016 Cell refractive index for cell biology and disease diagnosis: past, present and future *Lab Chip* **16** 634-44

[56] Bendib S and Bendib C 2018 Photonic crystals for malaria detection *J Biosens. Bioelectron 9 3 1000257*

[57] Shafkat A, Rashed A N Z, El-Hageen H M and Alatwi A M 2021 Design and analysis of a single elliptical channel photonic crystal fiber sensor for potential malaria detection *Journal of Sol-Gel Science and Technology* **98** 202-211

[58] Zaky Z A, Sharma A, Alamri S and Aly A H 2021 Theoretical evaluation of the refractive index sensing capability using the coupling of Tamm–Fano resonance in one-dimensional photonic crystals *Appl. Nanosci.* https://doi.org/10.1007/s13204-021-01965-7

[59] Barker Jr A and Ilegems M 1973 Infrared lattice vibrations and free-electron dispersion in GaN *Phys. Rev. B* **7** 743

[60] Yeh P 1988 *Optical waves in layered media* vol 95: Wiley New York)

[61] Skorobogatiy M and Yang J 2009 *Fundamentals of photonic crystal guiding*: Cambridge university press)

[62] Macleod H A 2017 *Thin-film optical filters*: CRC press)

[63] Zaky Z A and Aly A H 2020 Theoretical Study of a Tunable Low-Temperature Photonic Crystal Sensor Using Dielectric-Superconductor Nanocomposite Layers *J. Supercond. Nov. Magn* **33** 2983–2990

[64] Ayyanar N, Raja G T, Sharma M and Kumar D S 2018 Photonic crystal fiber-based refractive index sensor for early detection of cancer *IEEE Sens. J.* **18** 7093-7099

[65] Panda A and Devi P P 2020 Photonic crystal biosensor for refractive index based cancerous cell detection *Opt. Fiber Technol.* **54** 102123

[66] Zaky Z A, Ahmed A M, Shalaby A S and Aly A H 2020 Refractive index gas sensor based on the Tamm state in a one-dimensional photonic crystal: Theoretical optimisation *Sci. Rep.* **10** 9736

[67] Zeng C, Luo C, Hao L and Xie Y 2014 The research on magnetic tunable characteristics of photonic crystal defect localized modes with a defect layer of nanoparticle *Chin. Opt. Lett.* **12**

[68] Klimov V V, Pavlov A A, Treshin I V and Zabkov I V 2017 Fano resonances in a photonic crystal covered with a perforated gold film and its application to bio-sensing *J. Phys. D-Appl. Phys.* **50** 285101

[69] Mohammed N A, Hamed M M, Khalaf A A and EL-Rabaie S 2020 Malaria biosensors with ultra-sensitivity and quality factor based on cavity photonic crystal designs *The European Physical Journal Plus* **135** 1-22

[70] Gao Y, Dong P and Shi Y 2020 Suspended slotted photonic crystal cavities for high-sensitivity refractive index sensing *Opt. Exp.* **28** 12272-8

V-Trough Optimization for a Multipurpose Integrated Solar Energy Project in Helwan of Egypt

Yasser A Abdel-Hadi[1,*], Mohamed Sabry[1], Ahmed Ghitas[1], Harjit Singh[2] and David Redpath[2]

[1] PV Unit, Solar and Space Research Department, National Research Institute of Astronomy and Geophysics (NRIAG), Cairo, Egypt.
[2] College of Engineering, Design and Physical Sciences, Brunel University, London, UK

* E-mail: yasser_hadi@yahoo.com

Abstract. A solar concentrator of the type V-Trough is designed for a multipurpose integrated solar energy project in Helwan of Egypt. A suitable place in the NRIAG is selected. A concentration ratio matching the stability of the unconventional cooling technique for PV module 2.3 is chosen. A new cooling system with heat pipes for the module is manufactured. The optimum angles of setting the system for a maximum performance during the year are calculated using a ray tracing technique. The optimum dimensions of the PV module are 100 cm × 35 cm with an output voltage of 11.4 V and an output current of 4.29 A. The results show that establishing a solar thermal PV project with these conditions is promising for providing energy for off-grid local multipurpose integrated solar energy projects in that area.

Keywords: Solar Energy Concentration, Ray tracing, Electrical and Thermal PV module.

1. Introduction

An increased interest in the studies concerning hybrid Photovoltaic/Thermal (PV/T)solar energy systems for maximizing total output of the solar energy utilization [1], [2]. These studies revealed that such PV/T collectors have high overall efficiency than the traditional stand-alone PV and solar thermal systems operating separately.

A solar concentrator can increase the collected solar intensity by directing the incoming light rays onto a small collector area, enabling to operate under higher input solar power, hence producing higher output. Solar concentrators can share effectively in increasing the collected solar energy by concentrating such irradiation for effective thermal, electrical or hybrid energy conversion at affordable prices [3]. Solar concentrators are classified according to their optical device used (parabolic dish or trough, etc.), by their concentration ratio (low, medium and high), or by the type of their formed image (point focus, line focus, etc.) [4].

To reduce total system cost, non-imaging solar concentrators can efficiently use commercial reflective surfaces such as mirrors or high reflection films designed to collect extreme angular rays of incident solar radiation compared to just the axial rays collected by imaging systems [5]. Although their low optical tolerances, the development of non-imaging optics[6]enabled the development of line axis solar energy concentrating systems that can asymptotically approach the maximum geometric concentration of a light source for a particular angle of view. Non-imaging systems are more suited to Concentrating Solar Photovoltaic (CPV) systems as the distributed collected radiation across the collector provides a suitable uniformity of illumination required by solar cells to avoid shading, hence reduced performance[7].

V-trough solar concentrator is one of those non-imaging line axis optical systems, which offers an attractive option for integration to photovoltaic and thermal solar arrays. It is probably one of the least complex solar reflective concentrating systems to manufacture and operate with reasonable efficiency with zero or infrequent single axis adjustments, monthly or seasonal [8]. V-trough concentrators can also collect a reasonable amount of incident diffuse solar radiation at its aperture.

Many V-trough systems have been built up and tested all over the world with different concentration rations depending on their chosen orientation and tracking strategy. Reported concentrations ratios of as high as 4X and as low as 2X, and their collection efficiency have been varied accordingly. It has been mentioned that [6] the collected percentage of diffuse radiation incident on the concentrator aperture is the reciprocal of the concentration ratio. Optimizing the V-trough concentration ratio is then subject to the weather conditions and the percentage of diffused radiation intensity to that of direct.

The V-trough system under investigation has a chosen geometrical concentration ration of 2.3X and the collector is a PV/T novel hybrid system that maximizes its electrical output by cooling the PV using heat pipes and collect such thermal energy by means of active water-cooling sub-system. In this work, we only show the predicted V-trough optical performance in two different modes of orientations; yearly fixed and monthly tracking. Measured PV/T performance and the collection efficiency of the whole system will be published elsewhere.

2. Methodology

According to the topography of the area of the National Research Institute of Astronomy and Geophysics (NRIAG) and the distribution of the already existing buildings, the best location for setting the system was in the locations of latitude φ = 29.863526° N (29° 51′ 48.6936″ N) and longitude λ = 31.342028° E (31° 20′ 31.3008″ E). Figure 1 shows the sketch diagram of the location of the system beside the new building of the National Research Institute of Astronomy and Geophysics (NRIAG).

Figure 1. The sketch diagram of the location of the system besides the new building of the National Research Institute of Astronomy and Geophysics (NRIAG).

The system of V-Trough concentrator is supposed to be set at the position shown in figure 1 south-faced. There are two cases to optimize its performance; either to set it stationary with an optimum tilt angle or to track it seasonally or monthly according to the sun position during the year. For both cases, it is necessary to calculate the position of the sun during the year for the operation time of the system. Providing that the high operation hours of the system are during the day and especially around the noon time, we calculated the limits of the sun positions for the system location. It is obvious that the sun altitude reaches its higher value at noon. Table 1 shows the sun altitudes at noon for the cardinal days of the annual seasons.

Table 1. Sun altitudes at noon on the cardinal days of the annual seasons.

Date	Noon Time	Sun Declination	Sun Altitude at Noon
Winter Solstice (21st December)	11:56:22	-23.44°	36.48°
Spring Equinox (21st March)	12:01:37	0.4°	60.31°
Summer Solstice (21st June)	11:56:22	23.44°	83.34°
Autumn Equinox (21st September)	11:47:34	0.54°	60.45°

The limits of sun elevation angles in both winter and summer solstices are calculated to be 36.48° and 83.34° respectively. Therefore, the rays entering the V-Trough concentrator will be subtended between those angles. The optimum tilt angle for monthly variation has been statistically calculated by three specific anisotropic models (Tamps–Coulson, Perez and Bugler) at Helwan site [9]. Optimum tilt angle values measured from the zenith based on geometric factor are represented in Table 2. The experimental verification was presented based on weekly measurements obtained at a multi-position test facility[10].

Table 2. The predicted and experimental values of the monthly optimum tilt angle for the Helwan [9].

	Jan.	Feb.	Mar.	Apr.	May	June	July	Aug.	Sep.	Oct.	Nov.	Dec.
Predicted	55	45	30	15	5	5	5	15	25	40	50	55
Experiment	51	48	33	21	4	4	7	20	32	48	53	55

Figure 2 shows the Schematic diagram of the V-Trough concentrator which has to be analyzed to match the setting conditions at that location, where W_a is the width of the aperture, W_r is the width of the absorber and L is the length of segment which is the reflector side of the V-Trough.

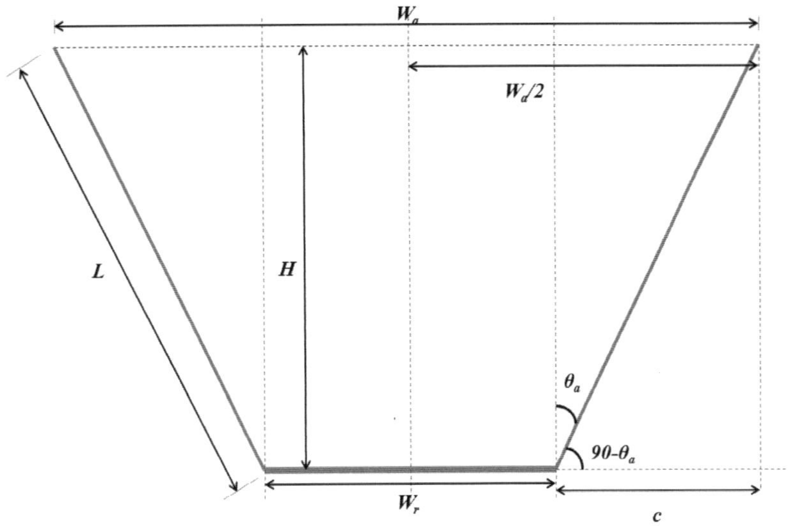

Figure 2. Schematic diagram of the V-Trough analysis.

If the acceptance angle is θ_a, then from figure2, we can derive the formulas from which we can calculate the dimensions of the V-Trough from the Equations (1to 3).

$$W_a = W_r + 2c \tag{1}$$

$$c = \frac{(W_a - W_r)}{2} \tag{2}$$

$$tan\theta_a = \frac{c}{H} \tag{3}$$

Then, the height of the V-trough can be calculated from Equation (4).

$$H = \frac{c}{tan\theta_a} = \frac{(\frac{W_a - W_r}{2})}{tan\theta_a} \tag{4}$$

Accordingly, the length of the trough segment can be calculated from Equation (5).

$$L = \frac{H}{cos\theta_a} \tag{5}$$

The goal is to use the ray tracing technique to find out the optimum design to redirect the solar rays within this angular range into the absorber. The tilting of the V-Trough concentrator (the direction of the normal of the absorber) is chosen to be the mean angle between 36.48° and 83.34°, which is56.38°. The absorber is a solar module of dimensions 100 cm × 35 cm shown in figure3. So, the width of the absorber is 35 cm. One of the aims of this experiment is to gain thermal energy besides electrical energy. Thus, the PV module is designed in a way to reduce solar cells temperature and collect such waste thermal energy to maximize total output. The photovoltaic module is consisting of five strings. Each string contains four monocrystalline silicon solar cells. The solar cells are fitted on top of an aluminum sheet for good thermal connection with a thin foil of EVA material for electrical isolation. The back aluminum sheets are grooved for fixing copper heat pipes, which are used for fast thermal dissipation. The top of the sheet strings is directly connected internally with an aluminum bulk copper tubes embedded for water flow. This new cooling technique is used for actively cooling the PV module when it is used under high concentration system.

According to the calculations of the whole system, the needed value of the concentration ration was taken to be 2.3. The reflectivity of the two segments of the concentrator is chosen to be 0.7 representing the normal low-cost commercial mirror. The half-acceptance angle of the trough θ_a will be 25.77° to match the needed concentration ratio. Accordingly, the width of the trough entrance (upper aperture) will be 80.5 cm, its height H will be 47.1 cm, while sidereal length of each one of its segments L will be 52.3 cm.

(a)

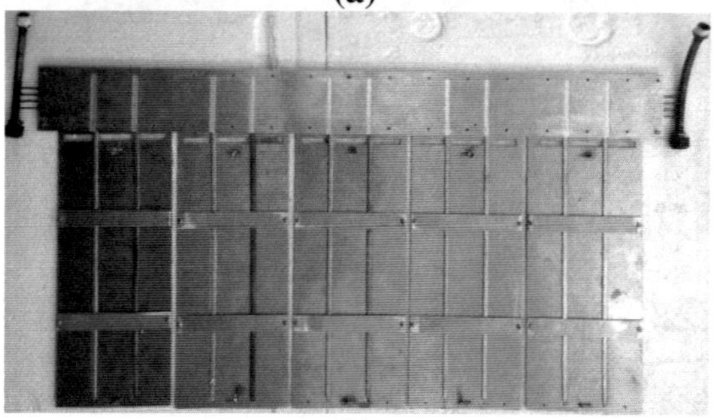

(b)

Figure 3. Monocrystalline silicon PV module connected with active cooler; (a) front and (b) back sides.

The trough is assumed to be set stationary with south facing tilted position with an angle equals to the mean altitude of the sun over the year at Helwan which is calculated to be 56.38°. The Trace Pro© software was used to simulate the distribution of the concentrated rays inside the trough. The incident solar irradiance of 750 W/m^2, which is supposed to be an average solar irradiance for a sunny day, was simulated for the V-Trough of the mentioned dimensions. Figure4 shows a schematic diagram of the reflections that the rays undergo inside the trough until they hit the absorber which is the PV-module.

Figure 4. A schematic diagram of the reflections that the rays undergo inside the trough until they hit the absorber which is the PV-module.

Also, it is important to simulate the distribution of the rays inside the trough after entering its upper aperture and undergoing their reflections to imagine their final distribution after the incidence on the PV-module. Figures5-8 show a simulation for the detected rays distributed on the PV module according to different angles of incidence representing the possible apparent positions of the sun in the sky. It is important to notice that the chosen stationary angle (56.38°) is the reference angle, to which the incident ray coming from the sun has an angle of 0°.

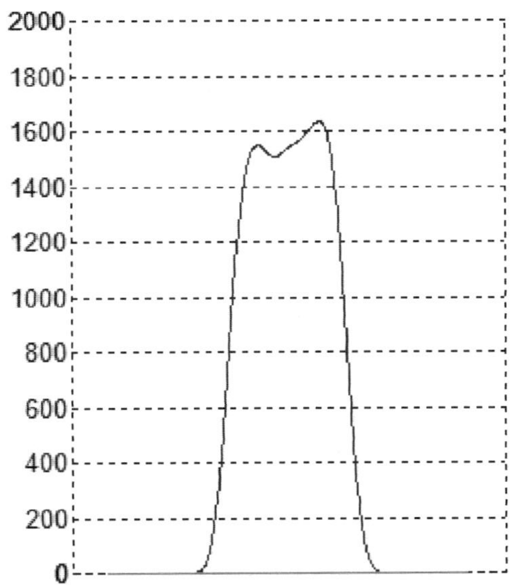

Figure 5. The distribution of the incident rays (W/m^2) on the PV module in the case of incident radiation of the same angle like the stationary facing angle of the V-trough (the difference in angles is 0°).

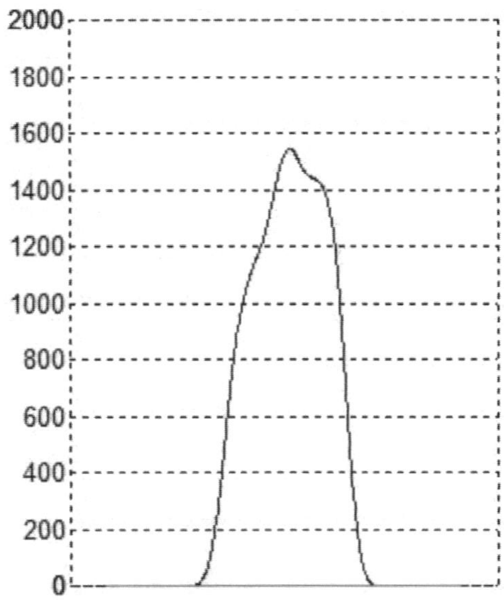

Figure 6. The distribution of the incident rays (W/m^2) on the PV module in the case of incident radiation 10° smaller than the stationary facing angle of the V-trough (the difference in angles is 5°).

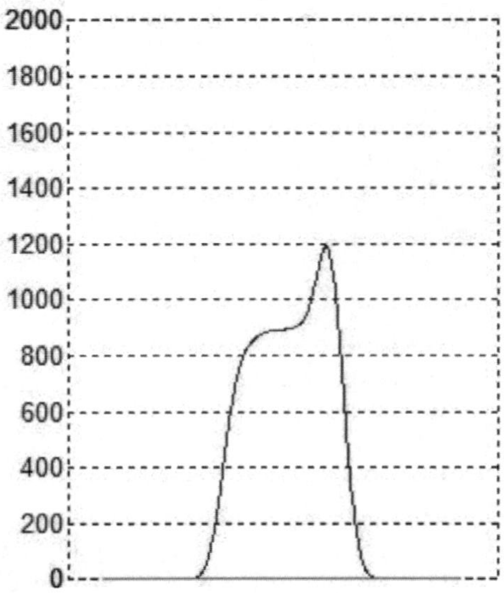

Figure 7. The distribution of the incident rays (W/m^2) on the PV module in the case of incident radiation 20° smaller than the stationary facing angle of the V-trough (the difference in angles is 10°).

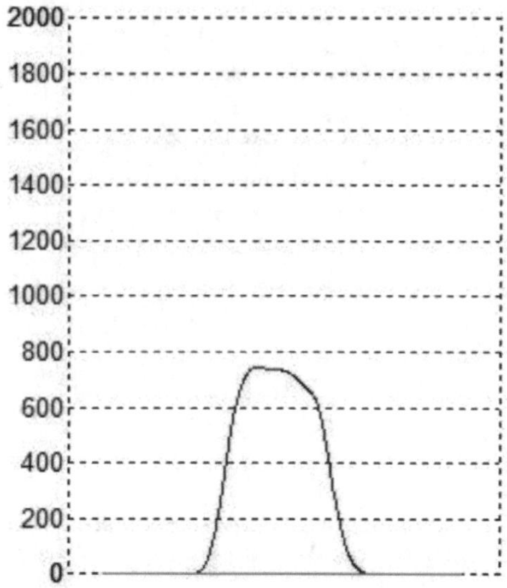

Figure 8. The distribution of the incident rays (W/m^2) on the PV module in the case of incident radiation 30° smaller than the stationary facing angle of the V-trough (the difference in angles is 20°).

It is important to calculate the solar radiation intensity delivered to the PV module by the V-Trough concentrator. Figure9shows these values against the differences between the tilting angle of the V-trough and the angle of the incidences of the solar radiation. Accordingly, we can calculate the intensity concentration ratio resulted by the concentrator against the tilting angle. Its values are represented in figure10. Figure 11 shows the angle of the incidence of the solar radiation on the cardinal days of the year seasons on the system tilted to the angle of 56° to the south during the operation time between 9:00 and 15:00. Accordingly, the concentration ratio of the V-Trough during the same operation according to the angle of the incidence of the solar radiation on the cardinal days of the year seasons on the system tilted to the angle of 56° to the south is shown in figure 12. Hence, the concentrated solar radiation intensity by the V-Trough delivered to the PV module during the operation time between 9:00 and 15:00 according to the angle of the incidence of the solar radiation on the cardinal days of the year seasons on the system tilted by angle of 56° is shown in figure 13.

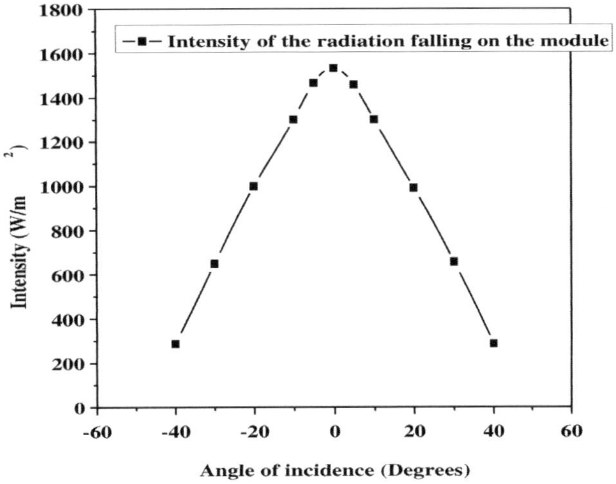

Figure 9. The concentrated solar radiation intensity received by the PV module against the differences between the tilting angle of the V-trough and the angle of the incidences of the solar radiation.

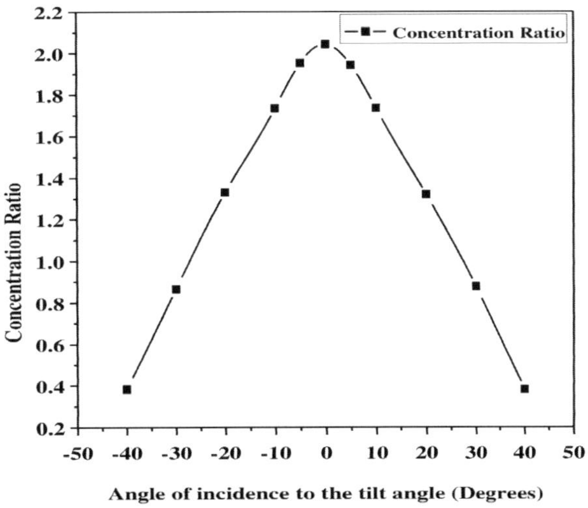

Figure 10. The concentration ratio of the V-Trough concentrator against the differences between the tilting angle of the V-trough and the angle of the incidences of the solar radiation.

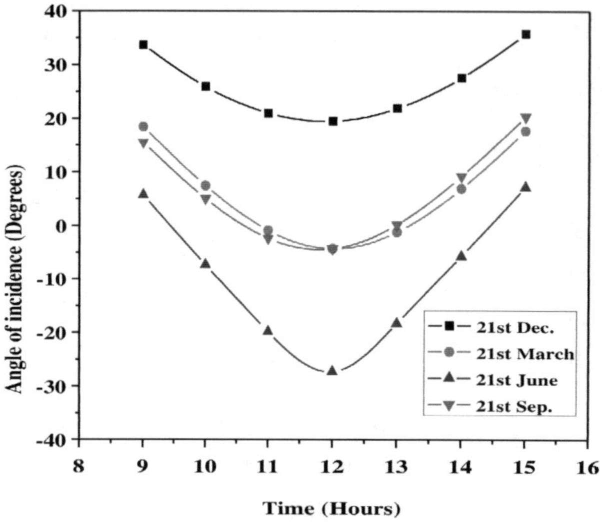

Figure 11. The angle of the incidence of the solar radiation on the cardinal days of the year seasons on the system tilted to the angle of 56° to the south during the operation time between 9:00 and 15:00.

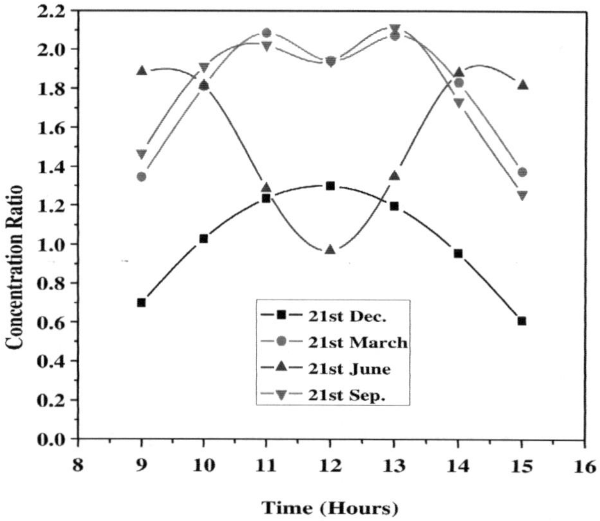

Figure 12. The concentration ratio of the V-Trough during the operation time between 9:00 and 15:00according to the angle of the incidence of the solar radiation on the cardinal days of the year seasons on the system tilted to the angle of 56° to the south.

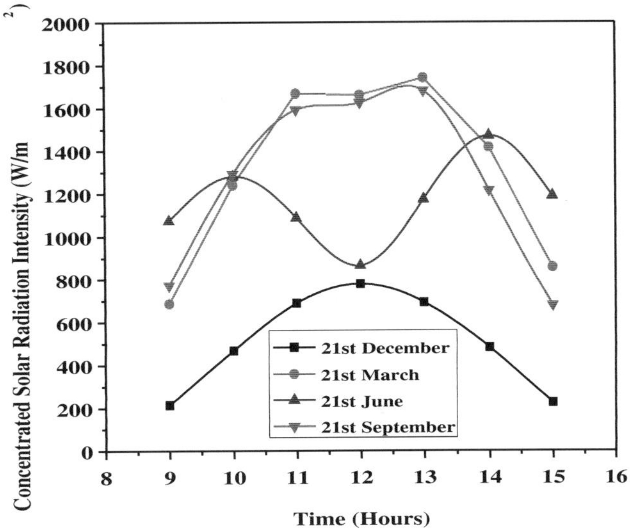

Figure 13. The concentrated solar radiation intensity by the V-Trough delivered to the PV module during the operation time between 9:00 and 15:00 according to the angle of the incidence of the solar radiation on the cardinal days of the year seasons on the system tilted by angle of 56°.

To increase the performance of the system, sometimes it is needed to apply the tracking mode. For such a two-dimensional system of concentrator, the optimum tracking mode is the seasonal tracking for the system which is set east-west facing the south. This means that the facing angle will be changed seasonally in order to collect the maximum possible solar radiation and to minimize the loss in the case of fixing the facing angle during the year. This will enhance the concentration ratio of the V-Trough to reach the upper limit of its geometrical value. Figure 14 shows the concentration ratio of the V-Trough during the operation time between 9:00 and 15:00 according to the angle of the incidence of the solar radiation on the cardinal days of the year seasons on the system tracked monthly. According to these values of concentration ratios, intensities of concentrated solar radiation shown in figure 15 can be attained.

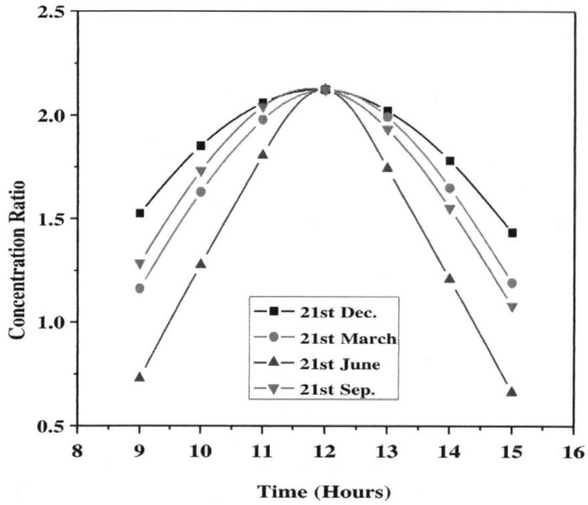

Figure 14. The concentration ratio of the V-Trough during the operation time between 9:00 and 15:00according to the angle of the incidence of the solar radiation on the cardinal days of the year seasons on the system tracked monthly.

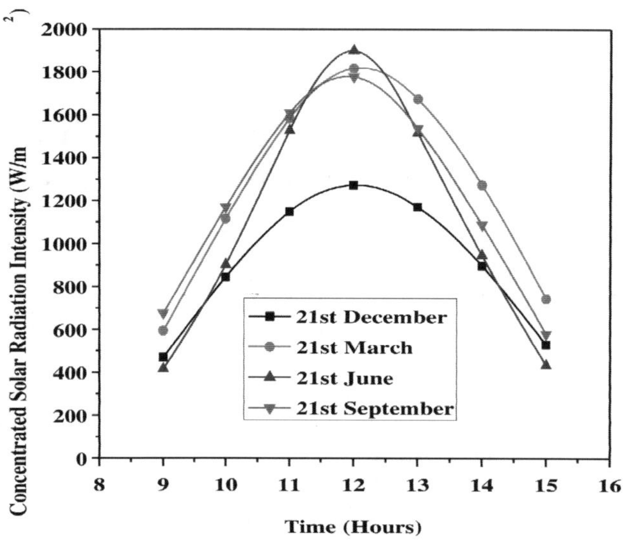

Figure 15. The concentrated solar radiation intensity by the V-Trough delivered to the PV module during the operation time between 9:00 and 15:00 according to the angle of the incidence of the solar radiation on the cardinal days of the year seasons on the system tracked monthly.

3. Predicted PV Output

The output field measurements of the PV module and separate strings are shown in figure 16. The measurements were carried out around the noon-time on March 15, 2021 at the project location. The intensity of the incident global solar radiation recorded at that time was 909 W/m^2. The tool used for the electrical measurements was PV- data logger. The total output module voltage and current in series connection are 11.4 V and 4.3 A respectively. Accordingly, the maximum output power attained by the module is 49.02 W. Hence, we can report the maximum efficiency of the module as

15%. Then, the output power gained by the module can be plotted against the tilting angle as shown in figure17.

Figure 16. Field measurements of the PV module.

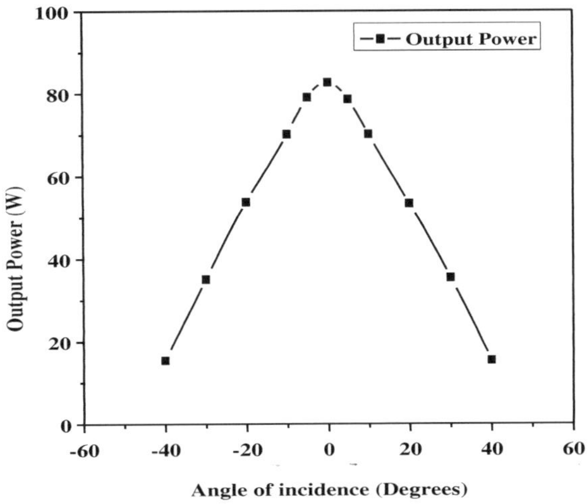

Figure 17. The output power of the system against the differences between the tilting angle of the V-trough and the angle of the incidences of the solar radiation.

From figure17, we can find that the power gained from the module during a spring day in the effective hours of the sun from 9:00 AM to 3:00 PM ranges between 40 and 80 W. The predicted output power obtained by the system in both cases (stationary tilted angle of 56° and monthly tracking by an optimum angle) can be represented in figures 18 and 19 respectively.

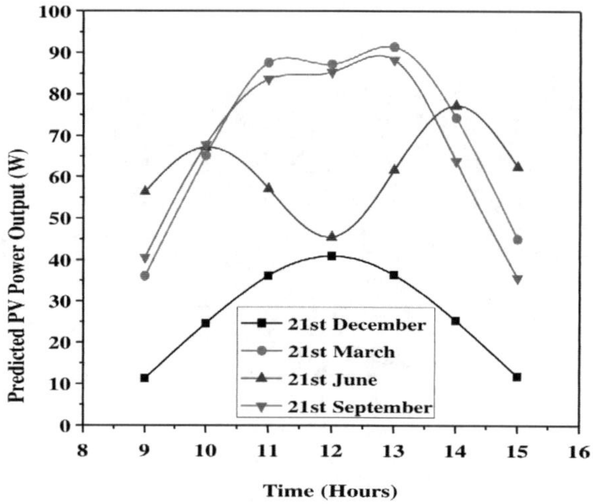

Figure 18. The predicted output power of the system during the operation time between 9:00 and 15:00 according to the angle of the incidence of the solar radiation on the cardinal days of the year seasons on the system tilted to the angle of 56° to the south.

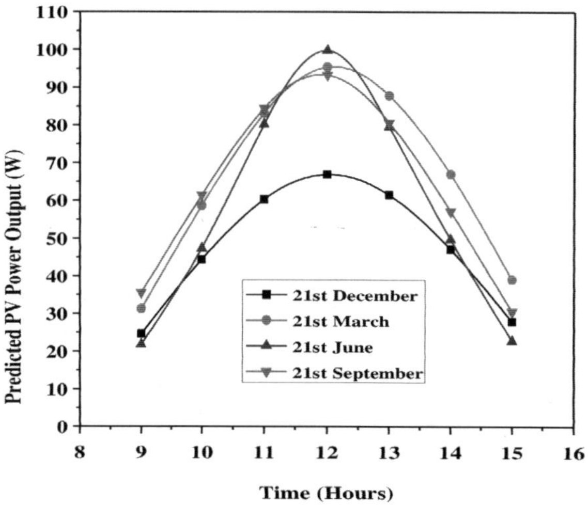

Figure 19. The predicted output power of the system during the operation time between 9:00 and 15:00 according to the angle of the incidence of the solar radiation on the cardinal days of the year seasons for the monthly tracked system.

4. Conclusions

In this work, we reported an evaluation and optimization of using a simply concentrated solar radiation for increasing the output power gained from a solar PV module in an arid location in Helwan (35 km south of Cairo, Egypt). Using position calculations and ray tracing technique, we optimized a simple PV module system receiving a concentrated solar radiation by a simple solar concentrator of a V-trough shape. By this system, we got more than twice the value of the incident global solar radiation concentrated on the PV module of an efficiency of about 15%. The system was chosen to be fixed on

a tilting angle calculated according to the location latitude and longitude. The distribution of the rays received by the module is ray-traced and accordingly calculated according to the difference between the sun's altitude and the fixed tilt angle. The maximum solar radiation power received by module was about 540 W which can be converted to about 83 W of electrical power.
The power gained from the module in such a system during a spring day in the effective hours of the sun from 9:00 AM to 3:00 PM ranges between 40 and 80 W. The output power is slightly dropped for the sun's altitude of 10° smaller or bigger than the fixed tilted angle of the system, while it drops dramatically (about the half of its maximum value) for the sun's altitude more than 20° smaller or bigger than that angle. Such power range can readily increase on the summer days on which the received global solar radiation has higher values.

Acknowledgment
Authors would like to thankfully acknowledge the UK IL Call 5 joint funding received for the NoNSTOP project from Egypt's Science, Technology and Innovation Funding Authority STIFA (project ID 30870) and UK's Newton Fund Institutional Links Programme (grant number 352029736)

References
[1] Van Helden WGJ, Van Zolingen RJC, Zondag HA 2004 *Progress in Photovoltaics: Research and Applications* **12** (6) 415–426
[2] Tripanagnostopoulos Y, Nousia TH, Souliotis M, Yianoulis P 2002 *Solar energy* **72** (3) 217-234
[3] Al-Najideen M, Al-Shidhani M, Min G 2019 *In AIP Conference Proceedings* **2149** (1) 030001
[4] Sharaf OZ, Orhan MF 2015 *Renewable and Sustainable Energy Reviews* **50** 1566-1633
[5] Welford WT, Winston R 1978 *Light and solar energy*
[6] Winston R 1974 *Solar Energy* **16** (2) 89–95
[7] Singh H, Sabry M, Redpath DAG 2016 *Solar Energy* **136** 421-427
[8] O'Gallagher JJ 2008 *Synthesis Lectures on Energy and the Environment: Technology*, Science, *and Society* **2** (1) 1-120
[9] Elminir HK, Ghitas A E, El-Hussainy F, Hamid R, Beheary MM, Abdel-Moneim KM 2006 *Energy conversion and management* **47** (5) 624-637
[10] Shaltout M, Hassan A, Ghitas A 1992 *PV Solar Energy Conference, Switzerland*

IOP Conf. Series: Materials Science and Engineering 1171 (2021) 012007 doi:10.1088/1757-899X/1171/1/012007

Study Of The Optimum Conditions And Parameters To Perform Simulation Of Low Temperature Capacitive Radio Frequency Argon Discharge

M Shihab[1,*], M G Elsheikh[2], T El-Ashram[2] and W M Moslem[2,3]

[1] Department of Physics, Faculty of Science, Tanta University, Tanta 31527, Egypt.

[2] Department of Physics, Faculty of Science, Port Said University, 42521 Port Said, Egypt.

[3]Centre for Theoretical Physics, The British University in Egypt (BUE), 43 El-Shorouk City, Cairo, Egypt.

*E-mail: mohammed.shihab@science.tanta.edu.eg

Abstract. Capacitively coupled plasma (CCP) is used extensively in plasma processing, such as, plasma etching, deposition, and sputtering. Particle-in-Cell (PIC) is used to explore the discharge dynamics microscopically. Considering all species in PIC simulation is challenging. Also, considering various atomic transitions, e.g., ionization, recombination, excitation, and deexcitation is not doable. The state of the art is to carry out fluid simulation. We compare between PIC and fluid simulation for radio frequency CCP to reveal the effect of various approximations as assuming constant temperature, simplifying Navier-Stokes equations in terms of particles mobility and diffusivity, and considering artificial boundary conditions at the electrodes. The fluid model predicts qualitatively PIC results in few minutes. For Argon discharge in geometrically symmetric CCP, the plasma bulk is quasineutral. Over the electrodes, two sheaths are built up due to the difference between ion and electron fluxes. The dynamics of two sheaths are out of phase. In the fluid model, chemical reactions and atomic processes can be considered. Argon metastable states density is maximum in the plasma bulk. In semi dark sheaths, metastable states concentration is small. The proposed fluid model could be used as a simulation platform to find the optimum conditions and to interpret experimental results.

1. Introduction

Capacitively coupled plasma reactors are characterized with simple geometry. They are simply a two planar, cylindrical or spherical electrodes. The discharge takes place between the electrodes. The potential difference between the electrodes may be driven by DC or/and Ac sources [1-3]. To generate high density plasmas in the bulk, the discharge is driven by radio frequency (RF) sources. The plasma bulk density increases by increasing the frequency [4]. Also, radio- frequency discharges allow the efficient processing of dielectric substrates as well as conductors. There are different configurations to allow control of the ion flux and ion energy over substrates. The two electrodes may be different in area, this kind of reactors called geometrically asymmetric capacitively coupled plasma reactors, where the sheath potential is inversely proportional to the electrode area [5]. Also, asymmetry between the sheaths could be achieved electrically when multiple harmonics sustain the discharge [6, 7]. Recently, magnetic fields are found to alter the discharge symmetry [8-10].

Argon is an inert gas. Its chemistry is simple. Therefore, it is used extensively in radio frequency capacitively coupled plasma reactors to understand the discharge dynamics experimentally and theoretically [11,12,13,14]. Also, in many industrial plasma applications, gases like O_2, N_2, CH_4, CF_4 are used with Argon to optimize the sputtered, deposited, or etched layers and profiles [15,16,17,18]. Of course, adding contaminations to Argon will change the chemical composition of plasma species, their concentrations, and the plasma density. In this manuscript we will model the Argon discharge. The model will be utilized as a simulation platform to add other gaseous in the future. It is not feasible to understand the plasma dynamics via solving the equation of motion of all plasma particles; the number of plasma particles is huge, and the motion of each charged particle affects the electric and magnetic fields distribution. Also, the general and exact solution of Boltzmann equation is not present [19]. Therefore, approximations have to be done to simplify the problem and at the same time to give acceptable physical description. The state of the art is to carry out a Particle-In-Cell (PIC) simulation [20, 21]. The equation of motion of superparticels is solved in a self-consistent way with Maxwell's equations. Each superparticle presents the physics of 10^3 to 10^6 of real particles. Although PIC simulation provides a microscopic description, the simulation is computationally expensive. One run may take few hours to few weeks depending on the underline problem and the available computer facilities. It is mandatory to find other methods to describe the plasma dynamics acceptably but in a proper time. One of these models is the lumped element circuits [22-24]. The discharge dynamics is calculated from the equivalent circuit. Resonances, harmonics generation, and power dissipations are well calculated in efficient time. Also, fluid models give a macroscopic description in a proper time. There are different approximations considering fluid models [25-29]. Here, we will assume drift-diffusion (DD) approximation at low pressure to investigate how far the results from PIC simulation. The fluid simulation assuming the DD approximation are in a good qualitative agreement with kinetically self-consistent simulation, however, the code is timely efficient. The long-time of PIC simulation is reduced to few minutes.

2. Fluid model

Fluid models provide a macroscopic description to the plasma species based on the averaged values of density, speed, and energy. The fluid model equations could be derived from the averaging of moments of Boltzmann equation. The equations set is closed with Maxwell's equations and assuming an equation of state. In typical capacitive coupled plasma, the gap size between the two electrodes is small compared to the area of the electrode; therefore, the problem could be simplified into 1D problem. If the distance between the electrodes is along z-axis, then the continuity equation for electrons, ions, and metastable states reads as:

$$\frac{\partial n_e}{\partial t} + \frac{\partial \Gamma_e}{\partial z} = K_{ei} n_n n_e + K_{mi} n_m n_e + K_{mm} n_m^2 , \tag{1}$$

$$\frac{\partial n_i}{\partial t} + \frac{\partial \Gamma_i}{\partial z} = K_{ei} n_n n_e + K_{mi} n_m n_e + K_{mm} n_m^2 , \tag{2}$$

And

$$\frac{\partial n_m}{\partial t}+\frac{\partial \Gamma_m}{\partial z}=K_{ex}n_n n_e-K_{mi}n_m n_e-K_{em}n_m n_e-K_r n_m n_e-2K_{mm}n_m^2-K_{2q}n_n n_m-K_{3q}\,n_n^2 n_m. \quad (3)$$

Where n_e, n_i, n_m, n_n are the density of electrons, Ar ions, Ar metastable states, and Ar neutral atoms, respectively. $\Gamma_j = n_j u_j$ is the species flux and u_j is the species speed where $j=e,\ I,\ m,\ n$ for electrons, ions, metastable, and neutral particles. Different atomic processes are considered as electron impact ionization of neutral species and metastable states, collision of metastable states, excitation of neutral species, and deexcitation of metastable states. The right hand side of equations 1, 2, and 3 are the gain and loss terms for each species with positive and negative signs, respectively. The rate coefficient of each process may be constant or temperature dependent as given in table 1 [30, 31].

The momentum conservation for low temperature plasma can be written as

$$m_e n_e\left(\frac{\partial u_e}{\partial t}+u_e\frac{\partial u_e}{\partial z}\right)=-en_eE-\nabla P_e-m_e n_e v_{en}u_e \quad (4)$$

$$m_i n_i\left(\frac{\partial u_i}{\partial t}+u_i\frac{\partial u_i}{\partial z}\right)=Zen_iE-\nabla P_i-m_i n_i v_{in}u_i \quad (5)$$

And

$$m_m n_m\left(\frac{\partial u_m}{\partial t}+u_m\frac{\partial u_m}{\partial z}\right)=-\nabla P_m-m_m n_m v_{mn}u_m \quad (6)$$

Where collisions with the neutral background is dominant. v_{jn} is the collision frequency of species j with the background gas. Collisions between electrons, ions, and metastable states are negligible. $P_j = n_j kT_j$ is the pressure of each species. The momentum conservation (NavierStokes equations) can be simplified utilizing the drift-diffusion (DD) approximation. Almost of the distance between the two electrodes are plasma bulk, where dependence on time and position is slowly varying, i.e., $\frac{\partial}{\partial t}\rightarrow 0$ and $\frac{\partial}{\partial z}\rightarrow 0$. Within the plasma bulk, the flux of species according to the DD approximation is given as

$$\Gamma_e=-n_e\mu_eE-D_e\frac{\partial n_e}{\partial x}, \quad (7)$$

$$\Gamma_i=n_i\mu_iE-D_i\frac{\partial n_i}{\partial x}, \quad (8)$$

and

$$\Gamma_m=-D_m\frac{\partial n_m}{\partial x}. \quad (9)$$

Table 1. Atomic processes considered in the simulation [30, 31]

Equation of Reaction		Rate of Reaction Coefficient
$e + Ar \rightarrow Ar^+ + 2e$	impact ionization	$K_{ei} = 1.253 \times 10^{-7} \exp(-18.618/T_e)\ cm^3/s$
$e + Ar \rightarrow Ar^* + e$	collisional excitation	$K_{ex} = 3.712 \times 10^{-8} \exp(-15.06/T_e)\ cm^3/s$
$e + Ar^* \rightarrow Ar^+ + 2e$	impact ionization	$K_{mi} = 2.05 \times 10^{-7} \exp(-4.95/T_e)\ cm^3/s$
$e + Ar^* \rightarrow Ar + e$	collisional deexcitation	$K_{em} = 1.818 \times 10^{-9} \exp(-2.14/T_e)\ cm^3/s$
$e + Ar^* \rightarrow Ar^r + e$	radiative deexcitation	$K_r = 2 \times 10^{-7}\ cm^3/s$
$Ar^* + Ar^* \rightarrow Ar^+ + Ar + e$	collisional ionization	$K_{mm} = 6.2 \times 10^{-10}\ cm^3/s$
$Ar^* + Ar \rightarrow 2Ar$	collisional deexcitation	$K_{2q} = 3.0 \times 10^{-15}\ cm^3/s$
$Ar^* + 2Ar \rightarrow Ar + Ar_2$	attachment	$K_{3q} = 1.1 \times 10^{-31}\ cm^6/s$

The fluxes are calculated based on the species mobility $\mu_j = e/m_j v_{jn}$ and diffusivity $D_j = T_e/m_j v_{jn}$ [32]. For electrons, the momentum transfer collisional frequency and the electron plasma frequency are larger than the driven radio-frequency, inertia of electrons may be neglected so that the mean velocities respond instantaneously to the electric field. Therefore, the DD approximation for electrons may be valid in the plasma sheath. On contrary, ions are heavier than electrons. Their mobility and diffusivity are small compared to electrons' mobility and diffusivity. Therefore, in the sheath, for ions we replace the instantaneous field E in equation 8 with the time averaged field; i.e.,$\bar{E} = \int_0^\tau E dt/\tau$. In the intermediate regime when the ion plasma frequency is comparable to the driven frequency, the DD approximation or its modification may be not accurate because ions respond partially to the instantaneous field.

The electric field E and the potential ϕ are calculated from

$$E = -\frac{\partial \phi}{\partial z} \tag{10}$$

and

$$\frac{\partial^2 \phi}{\partial z^2} = \frac{-e(n_i - n_e)}{\epsilon_o} . \tag{11}$$

In order to solve the equations' model numerically, the boundary conditions are as follows:
At left electrode

$$V_P = V_{DC} - V_{RF} \sin(\omega_{rf} t) \tag{12}$$

and at right electrode

$$V_G = V_{DC} - V_{RF} \sin(\omega_{rf} t + \pi). \tag{13}$$

The electron density and ion density between the electrodes, $0 < z < L$ are

$$n_{e,i} = n_0 \sin(\pi z/L). \tag{14}$$

At the electrodes the electron density is assumed to be inertial-less, so $n_{eP,G} = n_0 exp(eV_{P,G}/kT_e)$; the ion density is calculated from the balance between the ion flux and the electron flux to the electrodes and ensures the equality

$$V_{DC} = \sum_{k=1}^{\infty} \sum_{l=1}^{\infty} (n_e(k,l) - n_i(k,l)) \Delta z^2 \Delta t + V_f, \tag{15}$$

where k and l are the time and position indices on the time-space mesh, the area of each cell on the mesh is $\Delta z \Delta t$. Vf is the floating potential.

3. Results and discussion

IOP Conf. Series: Materials Science and Engineering 1171 (2021) 012007 doi:10.1088/1757-899X/1171/1/012007

An Argon gas with a pressure of 10 mTorr is considered. The distance between the two electrodes is 5 cm. The electron temperature is 1.5 eV and the ion temperature is 0.025 eV. Fluid simulation is iterated for 150 RF periods. The distance between the two electrodes is divided into 250 grids and each RF period is divided into 50 time steps. The traditional RF frequency of 13.65 MHz is used. The two electrodes are with the same area and a single frequency is driven. This is the geometrically and electrically symmetric discharge, where the dc self-bias of both sheaths is equal in magnitude and out of phase. Due to the high mobility and diffusivity of electrons with respect to ions, the electrode is always negatively biased with respect to the plasma.

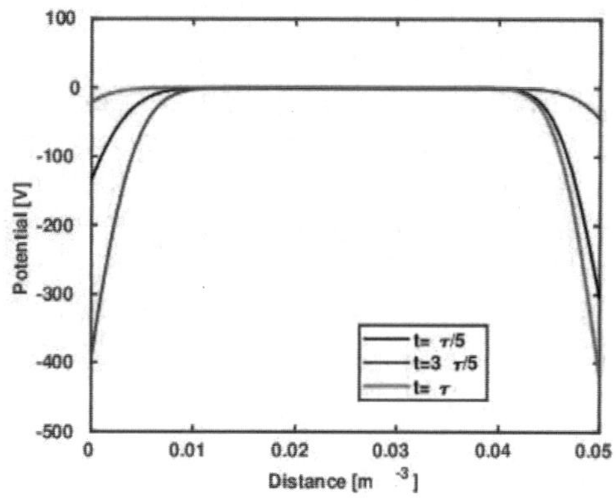

Figure 1. The potential between the electrodes at three phases, τ is the periodic time of the radio frequency voltage.

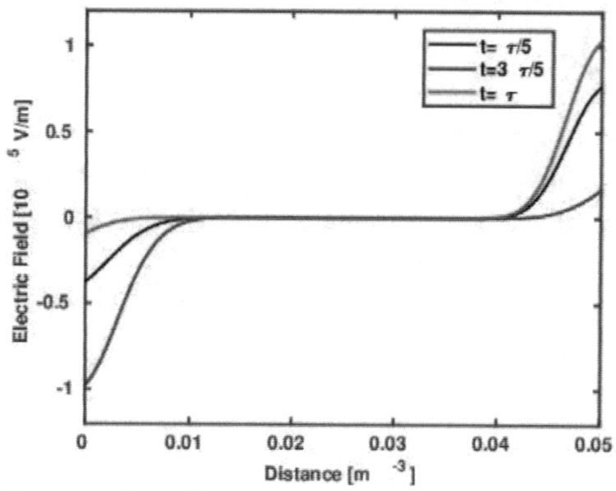

Figure 2. The electric field between the electrodes at three phases, τ is the periodic time of the radio frequency voltage.

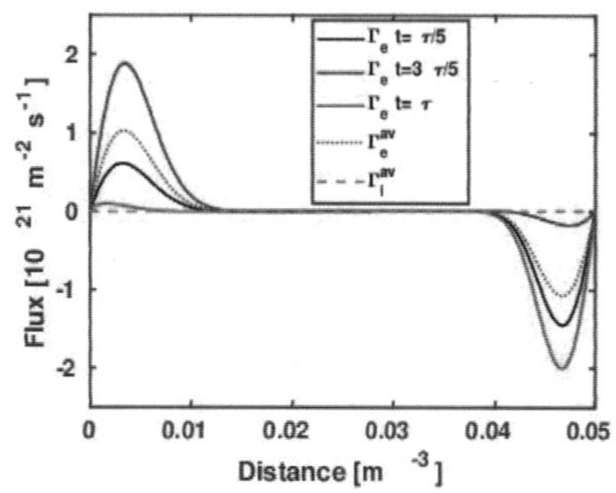

Figure 3. Electrons and ions fluxes between the two electrodes.

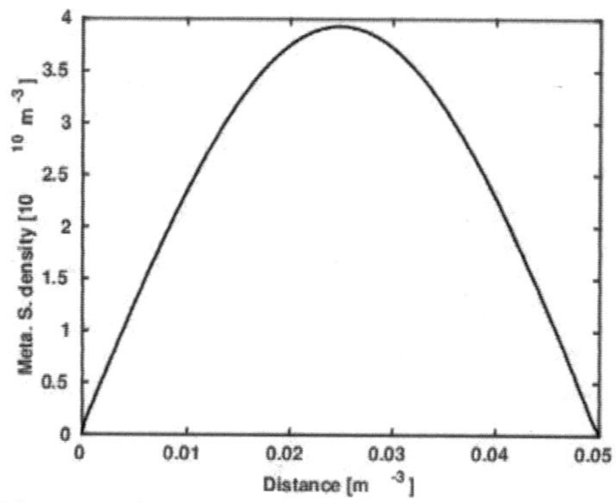

Figure 4. Time averaged metastable states density.

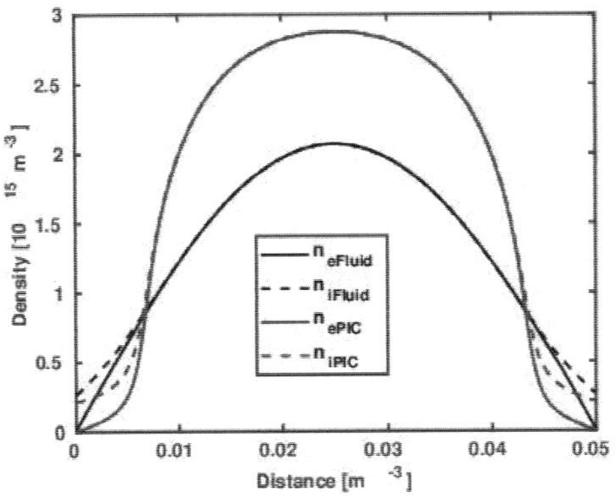

Figure 5. Time averaged densities between the two electrodes.

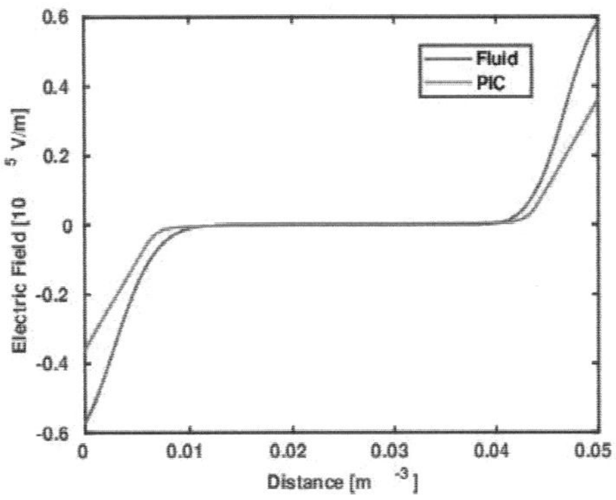

Figure 6. Time averaged field between the two electrodes.

Figure 1 displays the potential between the electrodes. The potential at the electrode is more negative compared to the plasma bulk. There is a phase difference π between the two sheaths. The collapse of a sheath is a companied with an expansion of the other sheath. Figure 1 provides the potential at three phases ($t = \tau$, $t = 3\tau/5$, and $t = \tau/5$) within the RF voltage signal. When the time equals the periodic time $t = \tau$, the left sheath is very small and its potential equals the floating potential, i.e., the potential due to the thermal motion of electrons. At $t = 3\tau/5$, the right sheath is collapsed with respect to the left sheath. Moreover, when most of the RF source voltage goes to a sheath, the potential on the opposite sheath is around the floating potential. The corresponding electric field between the two electrodes is shown in Figure 2. The sheath electric field accelerates positive ions to hit the substrate or the electrodes. From equations 7, electrons follow the applied field instantaneously. The electron fluxes at the three phases are shown in Figure 3. The time averaged electron flux is shown with a dotted line. The ion flux depends on the time averaged field and it is shown by the dashed line in Figure 3. To avoid misleading from Figure 3, the ion flux is not zero within the sheaths. But the time averaged electron flux is larger than the time averaged ion flux with the order of 10^3. This is a self-consistent with the results in Figure 1 and this is the reason behind the generation of the self-biased V_{DC}. The density of metastable states is shown in Figure 4, the maximum density of the metastable states are at the plasma bulk, where the brightness is also maximum. In order to check the validity of the DD approximation, the fluid results are compared with PIC results. In PIC simulation the distance between the two electrodes is discretized into 129 grids. The RF period is discretized into 600 time steps. To get convergence, the PIC simulation is iterated for 5000 RF periods. In fluid and PIC simulations, numerical instabilities are diminished. The time averaged density and field are shown in Figure 5 and Figure 6. This is a good qualitative agreement. As known in typical RF capacitive discharges, the plasma bulk is quasineutral. In the sheath, the ion density is larger than the electron density. The discrepancies between the fluid simulation and the PIC calculations could be minimized when the exact Navier-Stocks equations for ions are solved. Also, when the energy conservation and the variations of the particles mobility and diffusivity with temperature is considered. However, here, we got a macroscopic description in few minutes which is qualitatively agree with the computationally expensive PIC results. Our model can include different species of important gaseous as O_2, N_2, CH_4, CF_4, and, consequently, it may be used extensively in industrial applications. In addition, when the fluid model is coupled with a Monte Carlo scheme, the ion energy and the ion angular energy may be calculated in a hybrid manner [33] or fast self-consistent calculations could be carried out as in Ensemble-in-Space time (EST) model [34].

4. Conclusion

In this manuscript, a symmetric radio frequency capacitively coupled plasma has been simulated using Particle-in-Cell and a fluid model. Both models are in a qualitative agreement: The plasma bulk is quasineutral, two sheaths -where the electron density is smaller than the ion density- are created, and within sheaths a large field toward electrodes is generated. However, the fluid model is computationally efficient. In the fluid model, atomic processes are considered. The Metastable states of Argon are calculated between the electrodes. For high radio frequency regime; i.e., frequencies greater than 10 MHz, the assumption of constant electron temperature and the drift-diffusion approximation are valid. The proposed fluid model can be implemented by adding the chemistry of different species of important gaseous as O_2, N_2, CH_4, and CF_4. The proposed model can be used as a simulation platform to interpret and optimize etching, deposition, and sputtering experiments. Also, the fluid model could be coupled to a Monte Carlo scheme or to Ensemble-in-Space time model to calculate the particles distribution over the electrodes.

5. Acknowledgments

M. Shihab appreciates the great discussions with T. Mussenbrock concerning PIC results (Ruhr-University Bochum) and the financial support by the Academy of Scientific Research and Technology (ASRT), Egypt, Grant No 6742. ASRT is the 2nd affiliation of this research.

References

[1] Lieberman M A, Lichtenberg A J 2005 *Principles of Plasma Discharges and Materials Processing*, 2nd Edition (Wiley).
[2] Chabert P, Braithwaite N 2011 *Physics of Radio-Frequency Plasmas* (Cambridge: Cambridge University Press).
[3] Makabe T, Petrovic Z L 2014 *Plasma Electronics: Applications in Microelectronic Device Fabrication*, 2nd Edition (CRC Press).
[4] Shihab M, Mussenbrock T 2017 *Phys. of Plasmas* **24** 113510.
[5] Schulze J, Schüngel E, Czarnetzki U, Gebhardt M, Brinkmann R P, Mussenbrock T 2011 *Appl. Phys. Lett.* **98** 031501.
[6] Yang S, Zhang Y, Wang H, Jiang J C W 2017 *Plasma Process Polym.* **14** 1700087.
[7] Zhang Y-R, Hu Y-T, Gao F, Song Y-H and Wang Y-N 2018 *Plasma Sources Sci. Technol.* **27** 055003.
[8] Trieschmann J, Shihab M, Szeremley D, Elgendy A, Gallian S, Eremin D, Brinkmann R P, and Mussenbrock T 2013 *J. Phys. D: Appl. Phys.* **46** 084016.
[9] Saikia P, Bhuyan H, Escalona M, Favre M, Bora B, Kakati M, Wyndham E, Rawat R S, Schulze J 2018 *J. Appl. Phys.* **123** 183303.
[10] Oberberg M, Berger B, Buschheuer M, Engel D, Wölfel C, Eremin D, Lunze J, Brinkmann R P, Awakowicz P, Schulze J 2020 *Plasma Sources Sci. Technol.* **29** 075013.
[11] Hopwood J, Qian F 1995 *J. Appl. Phys.* **78** 758.
[12] Abdel-Fattah E 2013 *Phys. Lett. A* **377** 297**.**
[13] Kolobov V I, Arslanbekov R R, Levko Dmitry, Godyak V A 2020 *J. Phys. D: Appl. Phys.* **53** 25LT01.
[14] Klich M, Wilczek S, Janssen J F J, Brinkmann R P, Mussenbrock T, Triechmann J 2021 Plasma Sources Sci. Technol. **30** 065019.
[15] Prenzel M, Kortmann, von Keudell A, Nahif F, Schneider J M, Shihab M, Brinkmann R P 2012 *J. Phys. D: Appl. Phys.* **46** 084004.
[16] Rehman N U, Anjum Z, Masood A, Farooq M, Ahmad I, Zakaullah M 2013 Opt. Commun. **296** 72.
[17] Tsuji K, Hirokawa K 1992 *Appl. Sur. Sci.* **59** 31.
[18] Kikuchi H, Takahashi K, Mukaigawa S, Takaki K, Yukimura K 2021 Micromachines **12** 599.
[19] Inan U and Go lkowski M 2011 *Principles of Plasma Physics for Engineers and Scientists*, (Cambridge University Press).
[20] Wilczek S, Schulze J, Brinkmann R P, Z. Donkó, Trieschmann J, Mussenbrock T 2020 *J. Appl. Phys.* **127** 181101.
[21] Ranjan A, Ventzek P L G 2010 *Plasma Sources Sci. Technol.* **19** 015005.
[22] Shihab M 2018 *Physics Letters A* **382** 1609.
[23] Schmidt F, Trieschmann J, Gergs T, Mussenbrock T 2019 *J. Appl. Phys.* **125** 173106.
[24] Elgendy A 2021 *IEEE Trans. on Plasma Sci.* **49(2)** 711.
[25] Samir T, Liu Y, Zhao L-L, 2, Zhou Y-W 2017 *Chin. Phys. B* **11** 115201.

[26] Chang C-W, Davoudabadi M, Mashayek F 2010 *IEEE T PLASMA SCI* **38** 1603.
[27] Yarin A L, Rovagnati B, and Mashayek F, Matsoukas T 2006 *J. Appl. Phys.* **99** 064310.
[28] Gadoum A, Benyoucef D, and Lahoual M H L 2019 *Springer Nature Switzerland AG* 516522.
[29] Lu Y, Yan D-q, and Chen Y-s 2009 *J. Hydrodynamics* 21(6) 814.
[30] Lymberopoulos D P, Economou D J 1993 *J. Appl. Phys.* **73** 3668.
[31] Liu R, Liu Y, Jia W, Zhou Y 2017 *Phys. Plasmas* **24** 013517.
[32] Anders A 2005 *Surface and Coatings Tech.* **200** 1893.
[33] Diomede P, Longo S, Economou D J, and Capitelli M 2012 *J. Phys. D: Appl. Phys.* **45** 175204.
[34] Shihab M, Ziegler D, Brinkmann R P 2012 *J. Phys. D: Appl. Phys.* **45** 185202.

Low-temperature Hydrothermal Synthesis of Titanium Dioxide Nanoparticles for Photocatalytic Applications

M Abdelhamid Shahat[1,*], F M El-Hossary[2], Ahmed Ghitas[1], A M Abd El-Rahman[3,2] and A A Ebnalwaled[4]

[1] PV Unit, Solar and Space Research Department, National Research Institute of Astronomy and Geophysics (NRIAG), 11421 Helwan, Cairo, Egypt.
[2] Physics Department, Faculty of Science, Sohag University, Sohag, Egypt.
[3] King Abdulaziz University, Jeddah, KSA.
[4] Electronics & Nano Devices Lab, Physics Department, Faculty of Science, South Valley University, 83523 Qena, Egypt.

* E-mail: m.abdelhamid999@gmail.com

Abstract. Titanium dioxide nanoparticles (TiO_2 NPs) were synthesized by a surfactant-templated hydrothermal method at low temperature (75 ^{0}C) for photocatalytic applications. Investigation of structural, optical and photocatalytic properties of TiO_2 NPs were characterized by X-ray diffraction (XRD), scanning electron microscopy (SEM), high-resolution transmission electron microscopy (HR-TEM), FT-IR and UV–vis absorption spectroscopy. The structural analysis confirmed a polycrystalline tetragonal anatase TiO_2 phase with nanocrystalline structure and almost regular spherical shape. It has been found that the synthesized TiO_2 NPs have an energy bandgap value between 4.42 and 4.46 eV; achieving a photodegradation efficiency of methyl orange dye of 81% to 88 % under UV visible light.

Keywords: Photodegradation, TiO_2 NPs, Energy band gap, Hydrothermal Method

1. Introduction

Long-time ago, Organic dyes are considered as one of the most significant environmental and water pollutants. They are mostly non-biodegradable organic compounds (stable to light and oxidation) and have high toxicity to humans even at low concentrations [1]. Consequently, the large-scale disposal of toxic organics during wastewater purification considered a massive environmental challenge. Photocatalytic water treatment has been considered as a reliable technology, which aims to convert organic matter into inorganic compounds [2]. Metal oxides semiconductors (MOS) with good hydrophobicity and high photocatalytic activities have been effectively used in water treatment technology. They work on enhancing the oxidation of unmanageable pollutants suspended in water, through the generation of highly reactive hydroxyl (OH) and oxygen (O_2) free radicals [3]. In addition, metal oxides such as TiO_2, GO, rGO, CdS and Fe_2O_3 nanoparticles (NPs) are used as active materials for water purification process due to their ability to remove toxic elements on a large scale. [4–7]. Titanium dioxide nanoparticles (TiO_2 NPs) have attractive characteristics, such as high surface areas, synthesis simplicity, affordability, outstanding chemical stability and unique features (optical, electrical and magnetic). These properties are unique and suited for a wide spectrum of applications, such as PV cells, water splitting, electro-chromic devices, fillers, ceramics, coatings, pigment and photocatalysis [8]. In addition, it can be used in other applications for energy storage, sensors and optics [9–11]. Consequently, TiO_2 with its attractive properties and characteristics has found to be linked to fundamental and practical research. As known, TiO_2 is synthesized in different nanoforms, such as NPs films, nanorods, nanowires, and nanotubes, in which optimizing their properties can be adjusted by controlling preparation conditions [12]. For example, optimizing crystal size and phase structure lead to control of the TiO_2 NPs band gap for a specific kind of applications. Band gap energy (Eg) has a value of 3.0 and 3.1 eV for rutile and brookite structure, respectively. However, for anatase structure it was discovered to be more than 3.2 eV and hence UV light ($\lambda < 380$ nm) could be excited [13 – 15].

Bavykin et al. have prepared TiO_2 NPs in anatase, rutile and brookite structures using the hydrothermal process by producing and consecutively wrapping titanate nanosheets [16, 17]. Different physics and chemistry procedures were used to prepare TiO_2 NPs, for example solvothermal, sol-gel, sonochemical, hydrothermal, CVD, PVD and electro-deposition, etc. [18 – 24]. Herein, the surfactant-free hydrothermal process is employed as one of the soft solution techniques. It is a low-temperature process, used to synthesis various nanomaterials by creating a chemical reaction in water in a sealed pressure vessel [25]. As previously known, the degradation of the organic compound mechanism using TiO_2 is mainly based on the fast generation of electron-hole pairs by the irradiation with UV light. Then, it is followed by oxidation and reduction reactions [26]. TiO_2 works well as a heterogeneous photocatalyst in a dye polluted water environment and has the ability to oxidize donor molecules through photo-induced holes and reduce the adsorbed oxygen by electrons [27]. The following simple chemical equations show the oxidation and reduction process:

$$TiO_2 + hv \rightarrow TiO_2\,(h^+_{V.B} + e^-_{CB}) \quad (1)$$
$$h^+_{VB} + OH^- \rightarrow OH^{\bullet} \quad (2)$$
$$e^-_{CB} + O^-_2 \rightarrow O^{-\bullet}_2 \quad (3)$$
$$H_2O + O^{-\bullet}_2 \rightarrow OOH^{\bullet} + OH^- \quad (4)$$
$$2OOH^{\bullet} \rightarrow O_2 + H_2O_2 \quad (5)$$
$$H_2O_2 + O^{-\bullet}_2 \rightarrow OH^{\bullet} + OH^- + O_2 \quad (6)$$
$$OH^{\bullet} + O^{-\bullet}_2 + h^+_{V.B} + Pollutants \rightarrow CO_2\uparrow + H_2O \quad (7)$$

The current research work deals with the synthesis of TiO_2 NPs by surfactant-free hydrothermal method at low temperature. In this study, the hydrothermal reaction time is adjusted in order to control the crystal structure and optical properties of TiO_2 NPs, which in turn improve their photocatalytic efficiency for the degradation of organic dye in non-toxic forms.

2. Experimental Setup

2.1. Materials and Preparation method of TiO_2 NPs

As raw ingredients for the manufacture of TiO_2 NPs, TiO_2 powder (titanium powder, Ranbaxy,>5μm) were employed. The hydrothermal method of typical TiO_2 NPs is shown in Fig. 1. By placing 10 g of powder in a glass vial under a magnetic stirring (900 rpm) in 40 ml of distilled water for a 1h reaction period at room temperature. The mixture was sonicated for roughly 10 min then heat treated in a Teflon stainless steel autoclave at 75°C for 4h. The solution was able to cool down gradually to room temperature (RT) over 24h. The precipitate generated was filtered out, then washed with ethanol and water multiple times and calcinated for 3h at 500 °C. In general, the preceding process was employed to produce TiO_2 NPs at various periods of 8 and 12h of reaction.

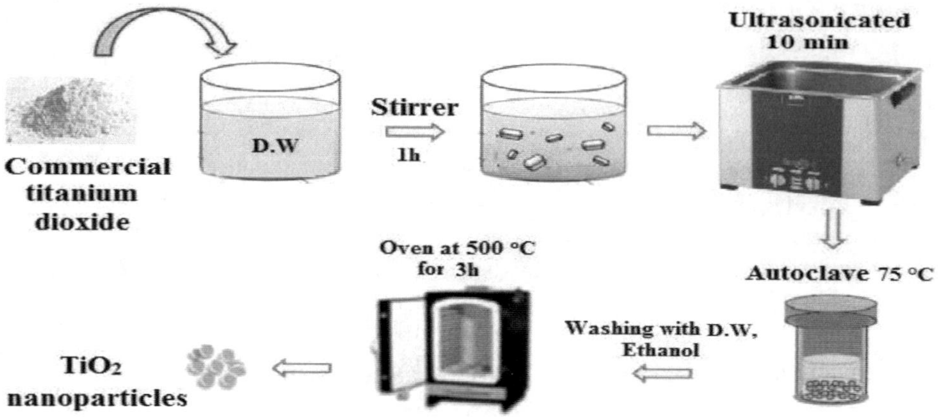

Figure 1. Preparation procedure of TiO_2 NPs by hydrothermal method.

2.2. Instrumentation and Characterization

The structure and phase purity of the synthesized samples were determined by X-ray powder diffraction (XRD, Philips diffractometer 1710) using Cu-Ka radiation (1. 5418 A^o). The morphology of the TiO_2 NPs was assessed by SEM (JOUL SEM model JSM – 5500 - Japan) with an accelerated voltage of 10 kV. Quantifying particle's shapes and features were characterized by HRTEM (JEOL, JEM- A 2100 – Japan) by applying an accelerating voltage of 200 kV. Furthermore, FT-IR (Jasco Model 4100 Japan) spectrum scanned from 4000 to 400cm^{-1} at RT with resolution of 4cm^{-1} have been evaluated for samples functional groups. In addition, under normal incidence wavelength ranging from 190-1100 nm, the optical characteristics of TiO_2 NP have been evaluated by a computerised SPECORD 200 PLUS spectrophotometer with 1nm step.

2.3. Photocatalytic activity

The photocatalytic activities of TiO_2 NPs were evaluated by the degradation level of the methyl orange (MO) dye. In order to achieve TiO_2 NPs with a catalyst concentration of 0.75 g/L, 75 mg of TiO_2 NPs was dissolved in 100ml of MO solution (50mg/L). The photocatalytic experiment was conveyed out in a beaker illuminated with a UV lamp (wavelength 264 nm). Preceding to photoirradiation, the nanoparticle suspension was stirred in a dark environment for 30 min to maintain adsorption/desorption equilibrium between the catalyst and MO dye molecules. The suspensions were collected, then centrifuged (4000 rpm, 15 min) at all irradiation intervals to separate the photocatalyst NPs. The UV-Vis Spectrophotometer (analytic Jena, Specord 200 plus, 190-1100 nm) was used to measure the concentrations of MO; the absorption was monitored at λmax = 464 nm. An alignment plot was carried out in relation to absorption and concentration, based on Beer-Lambert's law. Conducting experiments without catalysts was taken into consideration when the dye-catalyst solution was illuminated.

3. Results and Discussion

3.1. Phase formation and structural study

The phase identification and crystallinity of TiO_2 NPs samples were performed by X-ray diffraction. Fig. 2 shows the XRD patterns of the TiO_2 NPs hydrothermally prepared at various reaction times. Phase purity and crystallite size of the investigated samples were produced. The XRD spectrum of TiO_2 NPs exhibits tetragonal structure with clear peaks corresponding to the lattice planes of (101), (103), (200), (105), (213), (116), and (107). Table.1 presents d_{hkl} for the obtained TiO_2 NPs at different reaction times, which is in a good agreement with the standard XRD card [28].

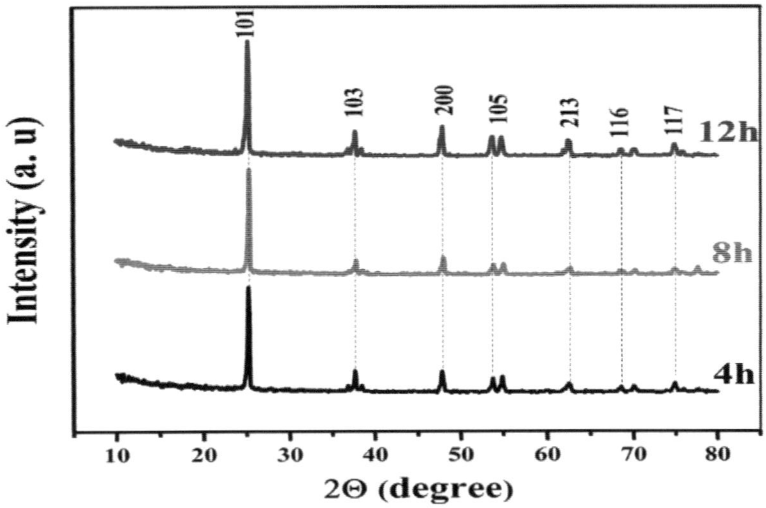

Figure 2. XRD spectra of TiO_2 NPs obtained at different reaction times.

Table 1. Comparison between d_{hkl} for the obtained TiO_2 NPs and d_{hkl} for the standard tetragonal TiO_2 crystals.

hkl	d_{hkl} (Å) Calculated			d_{hkl} (Å) for standard Tetragonal TiO_2
	4h	8h	12h	
(101)	3.52	3.52	3.52	3.52
(103)	2.39	2.38	2.38	2.37
(200)	1.90	1.89	1.89	1.88
(105)	1.70	1.70	1.70	1.70
(213)	1.48	1.48	1.48	1.48
(116)	1.36	1.36	1.36	1.36
(107)	1.26	1.27	1.26	1.26

Williamson-Hall analysis was applied to determine crystallite size (D) and microstrain (ε) of the investigated samples [29]. XRD peak broadening is concerned in these calculations. As illustrated from eq. (8), the total line broadening reflects two physical factors that consider the crystallite size (D) and strain (ε) effects:

$$\beta = \beta_D + \beta_\varepsilon \tag{8}$$

The Williamson-Hall equation is written as follows:

$$\Gamma cos\theta = \frac{0.94\lambda}{D} + 4\varepsilon sin\theta \tag{9}$$

The broadening factors of the TiO_2 NPs samples are directly calculated from drawing $\Gamma cos\theta$ versus sinθ as shown in Fig.3. Table (2) shows the crystallite size 31.48, 18.33 and 11.87 nm and their corresponding microstrain values 30, 33 and 34 x10^{-4} for TiO_2 NPs samples prepared for different reaction times of 4, 8 and 12h, respectively. The positive sign of the microstrain for TiO_2 indicates lattice expansion.

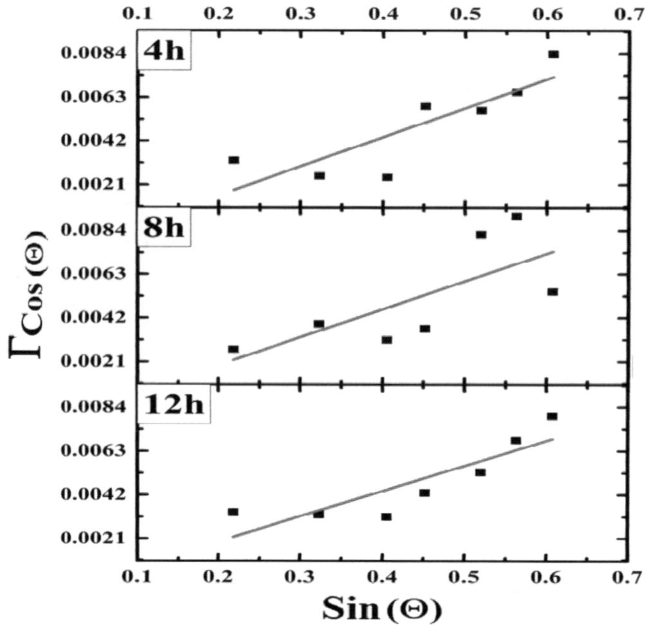

Figure 3. Williamson-Hall plot for TiO_2 NPs synthesized at different reaction times.

Table 2. Structural parameters of the obtained TiO_2 NPs at different reaction times.

Reaction time	Lattice-Parameters		Williamson-Hall	
	a (Å)	c (Å)	D size (nm)	Microstrain X (10^{-4})
4 h	3.77	9.48	31.47	30
8 h	3.77	9.48	18.33	33
12 h	3.78	9.48	11.86	34

3.2. Fourier transform infrared (FT-IR) spectroscopy

FT-IR is conducted as a spectroscopic technique to examine a wide range of organic materials. FT-IR spectra of the KBr sample pellets is analyzed in order to characterize the bond structure of TiO_2 NPs. All FT-IR spectra of TiO_2 NPs prepared at various reaction times (4, 8 and 12h) indicate three absorption bands as shown in Fig. 4. The FT-IR spectra of various TiO_2 NPs show the main absorption peaks with changes in the intensity of the absorption peak and peak width indicating the high influence of processing times on chemical bonds during the formation of TiO_2 NPs. The first peak appeared between 3700 and 3000 cm^{-1} confirms stretching hydroxyl (O-H), and represents water in moisture form. The second peak observed between 1700 and 1627 cm^{-1} is attributed to the stretching of titanium carboxylate. The third FT-IR spectrum peak that detected between 800 and 450 cm^{-1} corresponds to Ti-O stretching bands [30]. It has been previously stated that, preparing samples at relatively long hydrothermal reaction time led to detect strong absorption between 800 and 450 cm^{-1}. This was found to be attributed to the strong vibrations in the stretching Ti-O-Ti bonds [31, 32].

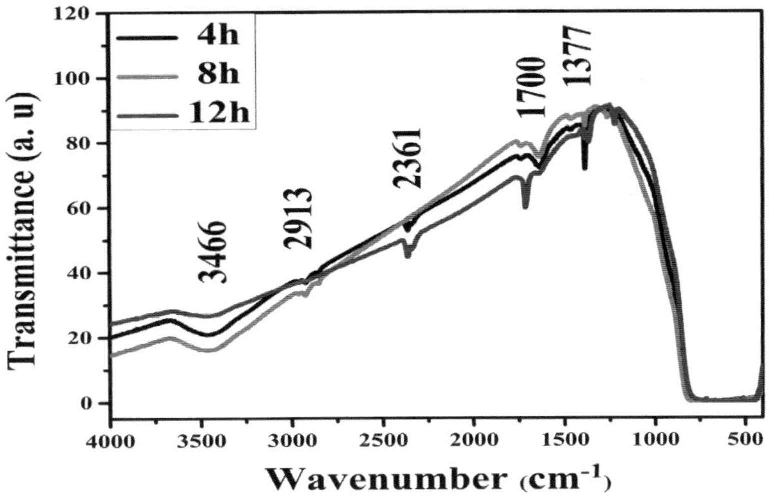

Figure 4. FT-IR spectra for TiO_2 NPs prepared at different reaction times.

3.3. Morphology

SEM and HR-TEM are used as microstructural characterization techniques in order to reinforce the phase identification analysis. Further, based on image analysis, particle shape and crystallite size of the TiO_2 NPs can be obtained. In HR-TEM analysis, the interference between the transmitted and the scattered beams leads to obtain a high-quality interference image. As shown from Fig. 5, the TiO_2 NPs are characterized with a semi-spherical particle form but with some variations in their particle size (≥ 50 nm) as the reaction time increases. It was observed that the sample, prepared at a 4 h reaction time, revealed TiO_2 NPs with a relatively smaller size than in samples prepared at 8 and 12 h. Similar results were obtained at relatively higher hydrothermal reaction temperatures [33 – 37]. Jun Ho Kim et al. succeeded in preparing TiO_2 NPs with uniform spherical shape by hydrothermal Teflon autoclave at 180 °C for 5 h, they found a relative increase in particle size owing to the sintering effect at different calcination temperatures (105- 700 °C) [33]. Sawanta S. Mali et al. prepared anatase TiO_2 with spherical shape nanoparticles using a hydrothermal reaction method at 160 °C for 4, 6 and 8 h [34]. Other research groups synthesized TiO_2 NPs with typical structural and morphological properties using the same method but with higher reaction temperatures and longer processing times; they increased the reaction temperature and time up to 240 °C and 36 h, respectively [35 – 37]. The formation mechanism of nanoparticles is agreed well with the previous nanoparticles growth mechanism. The hydrothermal growth mechanism for the TiO_2 NPs is summarized into four main stages, namely: (1) The dissolution process, in which the bonds of TiO_2 are broken in aqueous solution; (2) a thin-layer growth in a form of nanosheets; (3) nanoscale exfoliation process and finally (4) growing nanosheets with an increasing tendency of curling; leading to nanospheres' formation [38 – 41].

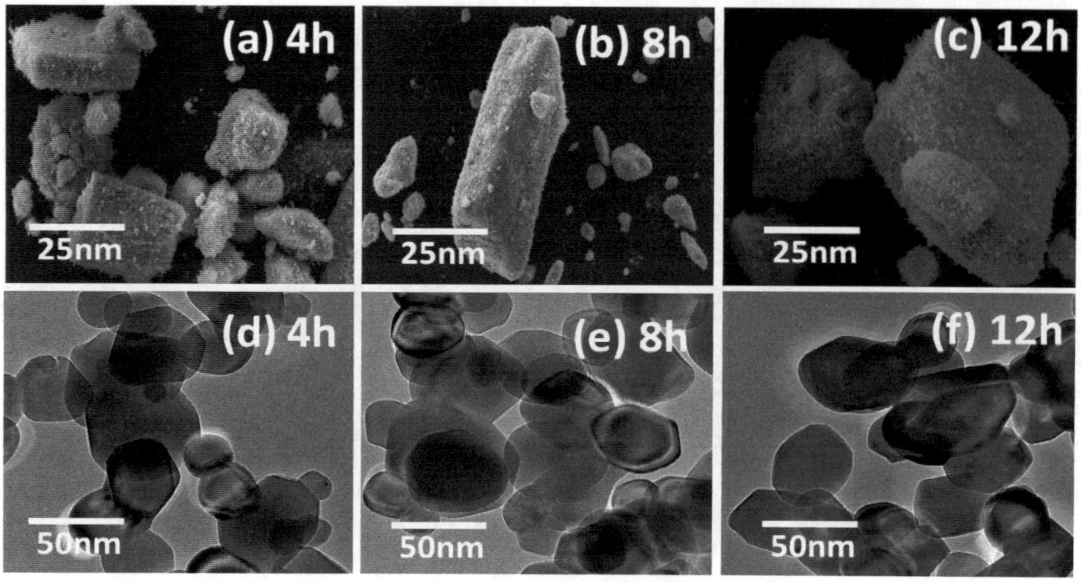

Figure 5.(a, b, c) Backscattered SEM images. (d, e, f) Interference HRTEM images for TiO_2 NPs prepared at different reaction times 4, 8 and 12h respectively.

3.4. Optical properties

One can obtain the desired absorption (maximum absorption) at a specific wavelength when light with different wavelengths interact with any material. UV-visible spectrophotometer is used to identify the absorption spectrum of TiO_2 NPs which are suspended in distilled water at room temperature. Fig. 6a shows the absorbance spectra of TiO_2 NPs over a spectral range between 200 nm and 800 nm. The absorption edge of all TiO_2 NPs samples is found to be equal and has a value of around 269 nm. This value is much lower compared to that of the bulk TiO_2 (374.9 nm). Almost the same behaviour has been observed by Hariharan et al. [42]. However, TiO_2 NPs doped by other materials of graphene/ Fe^{3+} and CdS found a relative higher absorption edge at 413 nm and at < 380 nm, respectively [43, 44]. Fig. 6b illustrates the transmission characteristics of TiO_2 NPs prepared at various reaction times. The TiO_2 NPs have a relatively higher transmission along the visible wavelengths.

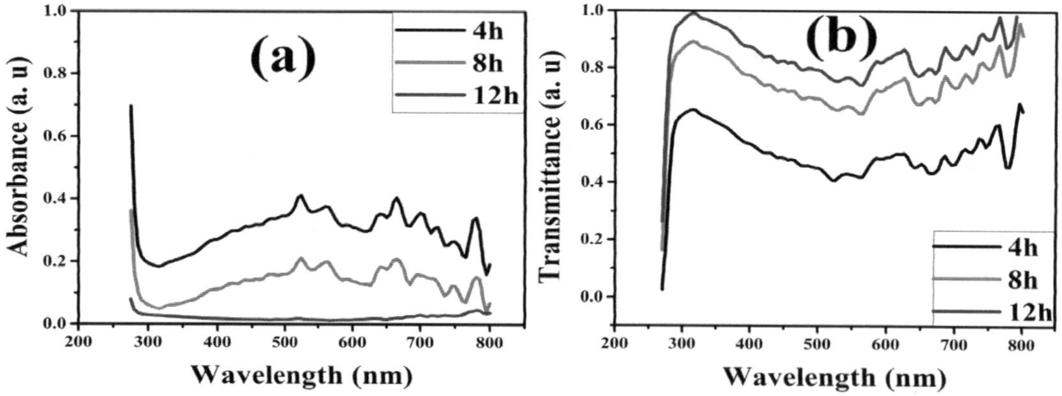

Figure 6a,b. UV-VIS absorption and transmission spectra for TiO_2 NPs prepared at different reaction time.

Tauc relationship is considered as a universal method used to determine the optical band gap for direct transition semiconductor materials [45 – 47]. Fig. 7a-c shows plots of $(\alpha h\nu)^2$ vs hν. The values of the optical band gap Eg were calculated for TiO_2 NPs samples prepared at different reaction times. Based on that, a limited change is obtained in the Eg values of TiO_2 NPs, indicating values of (4.46, 4.43 and 4.42 eV) for samples prepared at hydrothermal treatment times of 4, 8 and 12 h, respectively. Previously, the energy gap values of TiO_2 nanomaterials showed a high dependence on the preparation method as it achieved 3.85-4.32 eV in direct Eg values [48], and varied from 2.78 to 3.31 eV with the DC plasma sputtring method [49].

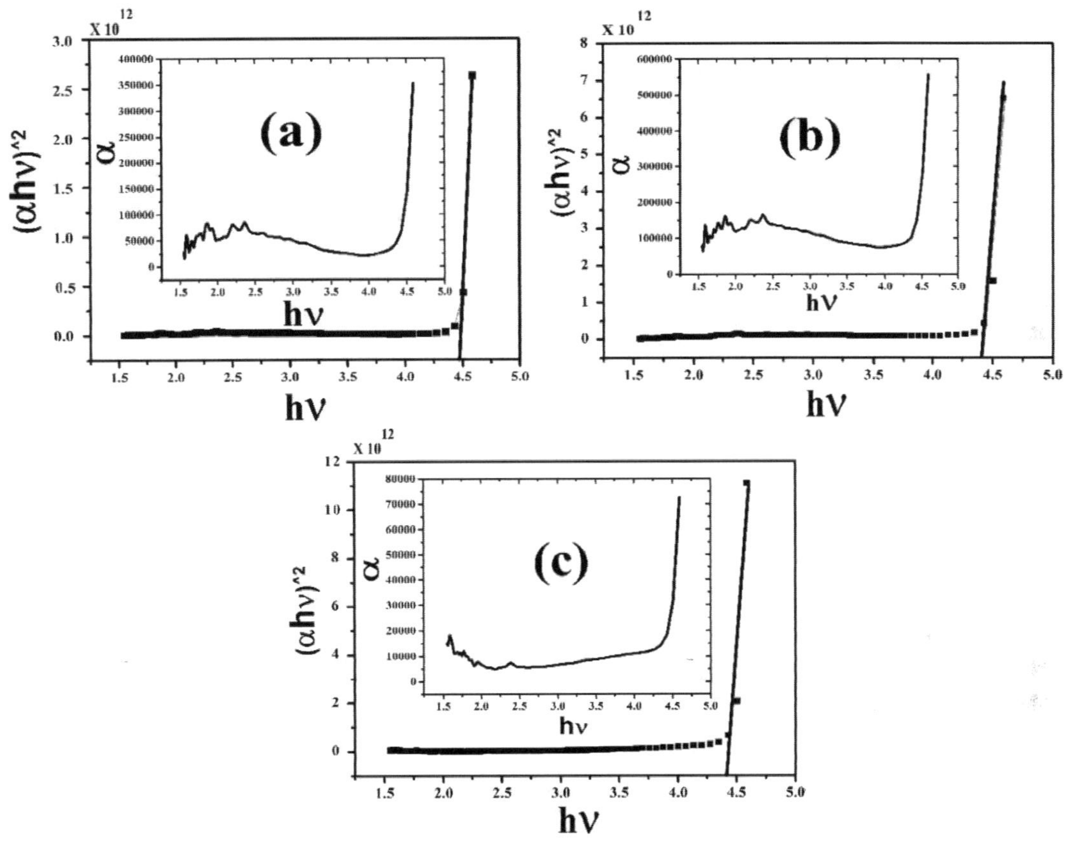

Figure 7a,b,c. The relation between $(\alpha h\nu)^2$ and hν, (a, b, c) for the obtained TiO_2 at different reaction times.

3.5. Water Treatment

Many organic compounds can be efficiently decomposed into the existence of photocatalytic metal oxide NPs. The degradation process of these organic compounds is realized by establishing a photooxidation reaction with the formation of surface hydroxyls (OH−) on the TiO_2 NPs. Electron–hole (e/h) pairs are generated in TiO_2 NPs during their irradiation with the UV light. To ensure the production of electron-hole pairs, the incident light should have energy higher or equal to the bandgap energy of the photocatalytic nanoparticles according to the photochemical reaction in eq.1 [50]. Photocatalytic materials with lower bandgap assure a fast generation of e/h pairs. This is accompanied by a low recombination possibility for the photogenerated charges carries produced in the water solution. The generated e/h pairs migrate to the interface and react with $O^{\bullet}_2$, OH^- that existed in TiO_2; leading to radicals formation consisting of hydroxyl cations ($\bullet OH$) and anions ($O^{\bullet}_2$).

$$(TiO_2) \cdots OH^- + h^+ \rightarrow \bullet OH \tag{10}$$

The hydroxyl radicals are the most effective oxidizing species which attack toxic organic pollutants and then produce nontoxic compounds (volatile CO_2 and water) [27]. Fig. 8a-c shows the UV-vis absorption spectra for MO dye solution after different time intervals treated with 75 mg of TiO_2 NPs at various reaction times of 4, 8 and 12h. The maximum absorption peaks are detected around 464 nm for all prepared solutions. The absorption peak intensity of TiO_2 NPs in the water-dye solution is observed to be gradually decreased with the increase of UV exposure time. Correspondingly, the MO dye decomposition is gradually decreased. Fig. 9 shows typical colors of MO dye solution mixed with TiO_2 NPs synthesized at a reaction time of 12h. As observed, orange to light yellow color degradation has detected with the gradual increase in UV exposure time. In Comparing the absorption spectra for MO dye solution treated with TiO_2 NPs, the highest photocatalytic activity is achieved with using TiO_2 NPs prepared at a reaction time of 12h. One can observe that the absorption peak for water treatment solution by TiO_2 NPs prepared at 12h has disappeared faster than for other solutions mixed with different TiO_2 NPs; leading to faster stopping for the dye decomposition. This is mostly attributed to structural changes of TiO_2 NPs, which is in particular found in the crystallinity improvement of TiO_2 NPs with the increase of hydrothermal reaction time [37].

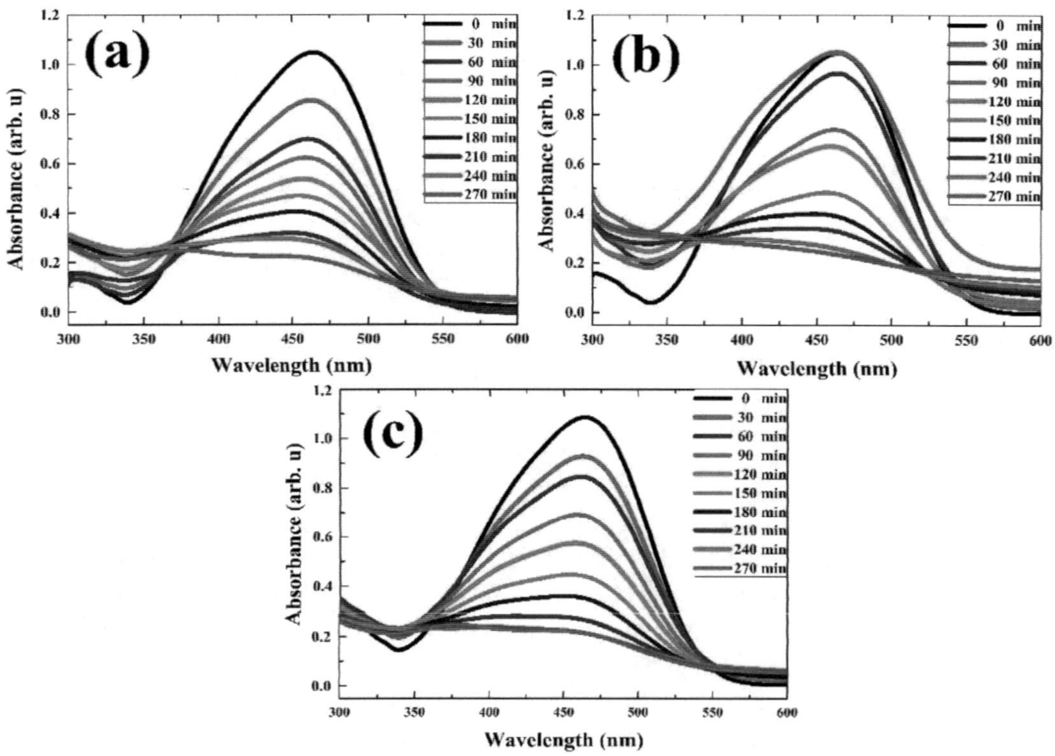

Figure 8a,b,c. Absorbance changes of MO after different irradiation times in the presence of TiO_2 photocatalyst under UV light illumination, (a, b, c) for the obtained TiO_2 at different reaction times.

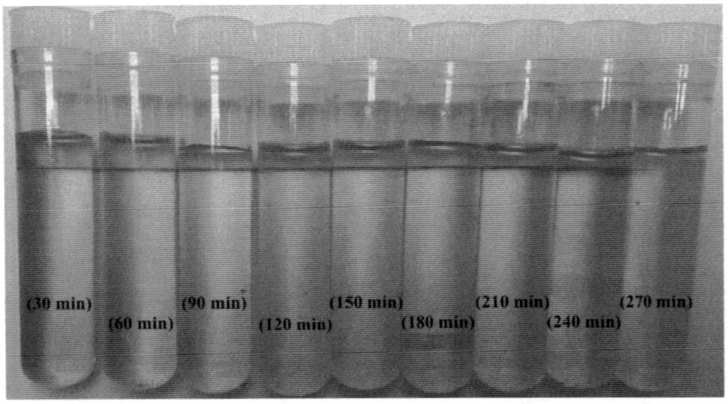

Figure 9. Typical colors matching experiment conducted on MO dye solution mixed with TiO_2 NPs synthesized at hydrothermal reaction time of 12h.

The photodegradation efficiency of the photocatalytic materials was calculated according to the simple formula of C / C_0, where C_0 is the photocatalytic initial concentration after equilibrium adsorption and C its concentration during the reaction time. Fig. 10a indicates a photocatalytic degradation for the MO dye in the presence of TiO_2 NPs. The results show a complete decomposition for the MO dye after 270 min under irradiation by UV-visible light. The anatase TiO_2 NPs with low energy band gaps, exhibit high photocatalytic activity owing to a rapid generation of e/h pairs [46]. This is owing to that the metal ions act as electron trap centers in the bandgap which leading to exceed carrier's lifetime and reduce carrier's recombination loss. These processes keep a high concentration of e/h pairs that directly exhibit high photocatalytic activity [46, 51]. The results that are presented in Fig. 10b explore that the organic materials cannot be decomposed without the photocatalyst or UV irradiation. The degradation efficiency was found to be gradually increased (81, 86 and 88%) after 270 min under UV irradiation with the increase of the hydrothermal reaction time of TiO_2 NPs (4, 8 and 12 h). The little change in the degradation efficiency is mostly ascribed to the gradual decrease in the values of the optical band gap (4.46, 4.43 and 4.42 eV). The photocatalytic efficiency of TiO_2 compared to that of ZnO, and GO in removing of methyl orange (MO) dye has been previously studied [52]. Other results indicated the best efficiency for TiO_2 in removing MO dyes with lower initial concentrations [53]. Further, the anatase TiO_2 as photocatalyst has been effectively used in the removal of MO and BPB dyes, with a degradation efficiency of 85.51% within 240 min [53].

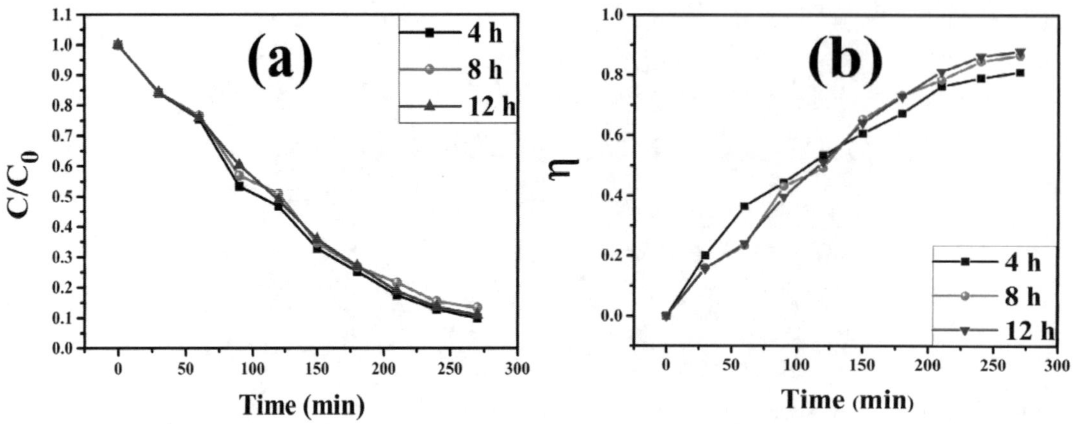

Figure 10a,b: a) Kinetic of the degradation of MO in the presence of UV, b) Degradation efficiency (η), as a function of the irradiation time for the obtained TiO_2 NPs prepared with different reaction times

4. Conclusion

The current work was accomplished with one of the most strategic metal oxide materials science research goals of finding an economic and efficient photodegradation system for water treatment applications. Photocatalyst TiO_2 NPs were synthesized by utilizing a free surfactant hydrothermal method at low temperature (75 ^{0}C) for various treatment times. The structural results confirm obtaining tetragonal TiO_2 NPs with the anatase phase. The crystallinity and surface morphology of the synthesized NPs are found to be influenced by changing the hydrothermal reaction time. The TiO_2 NPs have a spherical particle shape with a small crystallite size (31.4 -11.8 nm). Different reaction time process reveals that the bandgap decreased from 4.46 to 4.42 eV, which in turn leads to an increase in the degradation efficiency from 81 to 88% after 270 min under UV light irradiation which ensures good applications of TiO_2 NPs in water treatment.

References

[1] McCallum J E B, Madison S A, Alkan S, Depinto RL, Rojas Wahl RU 2000 *Environ Sci Technol* **34** (24) 5157–64
[2] Hadjltaief H B, Zina M B, Galvez M E, Da-Costa P 2016 *J Photochem Photobiol A Chem* **315** 25–33
[3] Chen D, Du Y, Zhu H, Deng Y 2014 *Appl Clay Sci* **87** 285–91
[4] El-Hossary FM, Ghitas A, El-Rahman AMA, Shahat MA, Fawey MH 2021 *Vacuum* 110158
[5] El-Hossary FM, Ghitas A, Abd El-Rahman AM, Ebnalwaled AA, Shahat MA, Fawey MH 2021 *J Mater Sci Mater Electron* 1–14
[6] El-Hossary F M, Ghitas A, Abd El-Rahman AM, Ebnalwaled AA, Shahat MA, Fawey MH 2020 *IOP Conference Series: Materials Science and Engineering* **956** (1)
[7] El-Hossary FM, Ghitas A, Abd El-Rahman AM, Ebnalwaled AA, Fawey MH, Shahat MA 2020 *IOP Conference Series: Materials Science and Engineering* **762** 012001
[8] Lamba R, Umar A, Mehta SK, Kansal SK 2015 *J Alloys Compd* **653** 327–33
[9] Osman DAM, Mustafa MA. 2015 *J Nanosci Nanoeng* **1** (4) 248–51
[10] Benazir A, Gomathi K, Aram S 2017 *J Environ Nanotechnol* **6** (1) 39–43
[11] Panda KK, Golari D, Venugopal A, Achary VMM, Phaomei G, Parinandi NL, et al 2017 *root bioassay system. Antioxidants* **6** (2) 35
[12] Behnajady MA, Modirshahla N, Fathi H 2006 J Hazard Mater **136** (3) 816–21
[13] Chen X, Mao SS. 2007 Chem Rev **107** (7) 2891–959
[14] Kuo H-L, Kuo C-Y, Liu C-H, Chao J-H, Lin C-H. 2007 *Catal Letters* **113** (1–2) 7–12

[15] Reyes-Coronado D, Rodríguez-Gattorno G, Espinosa-Pesqueira ME, Cab C, De Coss R d, Oskam G. 2008 *Nanotechnology* **19** (14) 145605
[16] Bavykin D V, Parmon VN, Lapkin AA, Walsh FC 2004 *J Mater Chem* **14** (22) 3370–7
[17] Peng T, Ray S, Veeravalli SS, Lalman JA, Arefi-Khonsari F 2018 Mater Res Bull **105** 104–13
[18] Manjunath K, Yadav LSR, Jayalakshmi T, Reddy V, Rajanaika H, Nagaraju G 2018 *J Mater Res Technol* **7** (1) 7–13
[19] Lin H, de Oliveira PW, Grobelsek I, Haettich A, Veith M 2010 *Zeitschrift für Anorg und Allg Chemie* **636** (11) 1947–54
[20] Manjunath K, Souza VS, Ramakrishnappa T, Nagaraju G, Scholten JD, Dupont J 2016 *Mater Res Express* **3** (11) 115904
[21] Manjunath K, Souza VS, Nagaraju G, Santos JML, Dupont J, Ramakrishnappa T 2016 *New J Chem* **40** (12) 10172–80
[22] Nagaraju G, Ravishankar TN, Manjunatha K, Sarkar S, Nagabhushana H, Goncalves R, et al 2013 *Mater Lett* **109** 27–30
[23] Nagaraju G, Manjunath K, Sarkar S, Gunter E, Teixeira SR, Dupont J 2015 *Int J Hydrogen Energy* **40** (36) 12209–16
[24] Krishnappa M, Souza VS, Ganganagappa N, Scholten JD, Teixeira SR, Dupont J, et al 2015 Chem Eur J **21** (49) 17624–30
[25] Adschiri T, Hakuta Y, Sue K, Arai K 2001 *J Nanoparticle Res* **3** (2–3) 227–35
[26] Hariharan D, Srinivasan K, Nehru LC 2017 *J Nanomedicine Res* **5** (6) 138–42
[27] Ramakrishnan VM, Natarajan M, Santhanam A, Asokan V, Velauthapillai D 2018 *Mater Res Bull* **97** 351–60
[28] Hanawalt JD, Rinn HW, Frevel LK. Chemical analysis by X-ray diffraction 1938 *Ind Eng Chem Anal Ed* **10** (9) 457–512
[29] Hamann TW, Martinson ABF, Elam JW, Pellin MJ, Hupp JT. 2008 *Adv Mater* **20** (8) 1560–4
[30] Dobrucka R 2017 *Iran J Pharm Res IJPR* **16** (2) 756
[31] Guo Y, He D, Xia S, Xie X, Gao X, Zhang Q 2012 *J Nanomater* 2012
[32] Kumar PSM, Francis AP, Devasena T 2014 *J Environ Nanotechnol* **3** (4) 78–85
[33] Kim JH, Noh BH, Lee G-D, Hong S-S 2005 *Korean J Chem Eng* **22** (3) 370–4
[34] Mali SS, Betty CA, Bhosale PN, Patil PS 2011 *CrystEngComm* **13** (21):6349–51
[35] Collazzo GC, Jahn SL, Carreño NL V, Foletto EL 2011 J Chem Eng **28** (2) 265–72
[36] Vijayalakshmi R, Rajendran V 2012 *Arch Appl Sci Res* **4** (2) 1183–90
[37] Kavitha M, Gopinathan C, Pandi P 2013 *Int J Adv Res Technol* **2** (4) 102–8
[38] Shahrezaei M, Habibzadeh S, Babaluo AA, Hosseinkhani H, Haghighi M, Hasanzadeh A, et al 2017 *J Exp Nanosci* **12** (1) 45–61
[39] Cui L, Hui KN, Hui KS, Lee SK, Zhou W, Wan ZP, et al 2012 *Mater Lett* **75** 175–8
[40] Chao C, Ren Z, Yin S, Xu G, Gong S, Yang X, et al 2014 *Adv Powder Technol* **25** (2) 745–51
[41] Weng L-Q, Song S-H, Hodgson S, Baker A, Yu J 2006 *J Eur Ceram Soc* **26** (8)1405–9
[42] Hariharan D, Christy AJ, Mayandi J, Nehru LC 2018 *Mater Lett* **222** 45–9
[43] Li W, Liu X, Li H 2015 *J Mater Chem A* 3 (29) 15214–24
[44] Xie Z, Liu X, Wang W, Liu C, Li Z, Zhang Z 2014 *Sci Technol Adv Mater* **15** (5) 55006
[45] Pankove JI 1975 *Courier Corporation*
[46] Ahmadi M, Ghasemi MR, Rafsanjani HH 2011 *J Mater Sci Eng* **5** (1) 87
[47] El-Hossary FM, Ghitas A, Abd El-Rahman AM, Ebnalwaled AA, Shahat MA 2020 *IOP Conference Series: Materials Science and Engineering* **956** 012003
[48] Rahman K H, Biswas S, & Kar A K 2018 AIP Conference Proceedings 1 (1953) 030022
[49] Abd El-Rahman A M, & Mohamed S H 2020 Thin Solid Films 698 137864
[50] Kontos A I, Arabatzis I M, Tsoukleris DS, Kontos AG, Bernard MC, Petrakis DE, et al 2015 *Catal Today* **101** (3–4) 275–81
[51] Ni M, Leung MKH, Leung DYC, Sumathy K 2007 *Renew Sustain Energy Rev* **11** (3) 401–25
[52] Raliya R, Avery C, Chakrabarti S, Biswas P 2017 *Appl Nanosci* **7** (5) 253–9
[53] Dhanalakshmi J, Padiyan DP 2017 *Mater Res Express* **4** (9) 95020

Comparative neutronic study for heterogeneous and homogeneous fuel assembly in a lead-cooled fast reactor

Mohammed Ibrahim[1], Amr Ibrahim[2], Moustafa Aziz[3], H A Saudi[4] and M Y Hassaan[5]

[1] School of Engineering and Applied Sciences (EAS), Nile University, Giza, Egypt.
[2] High Institute for Engineering and Technology at Al-Obour, Cairo, Egypt.
[3] Nuclear Regulatory Commission and Radiological, Nasr City, Cairo, Egypt.
[4] Physics Department, Faculty of Science, Al-Azhar University (Girls Branch), Cairo, Egypt.
[5] Physics Department, Faculty of Science, Al-Azhar University, Nasr City 11884,Cairo, Egypt.

E-mail: mibrahem@nu.edu.eg

Abstract. ALFRED (Advanced Lead-cooled Fast Reactor European Demonstrator) is a (300 MW_{th}) pool-type reactor with closed hexagonal Fuel Assemblies (FAs) which are divided into two radial zones (inner and outer core zones). In this work two three dimensional models (3D-heterogenous and 3D-homogeneous) for inner assembly (IA) and outer assembly (OA) have been designed using MCNPX transport code to simulate the neutronic behavior and to study the fuel performance and the effect of homogenization inside the fuel assemblies. The results of the present work for power distribution and fuel burn-up show good agreement between 3D-homogeneous and 3D-heterogeneous models.

1. Introduction

In 2002, fast spectrum reactor cooled by lead was considered as one of three systems of advanced fast spectrum nuclear reactors by the fourth Generation International Forum (GIF) [1]. Thus, fast spectrum reactors are the most efficient technology for effectively utilizing uranium resources because of the ability of using the uranium stored in spent nuclear fuel to be recycled while providing energy and consequently overs a realistic path forward for achieving energy sustainability consistent with environmental stewardship by introducing an advanced technology for good management of long lived minor actinides and nuclear fission products [2]. Among all fourth generation (Gen-IV) advanced innovative concepts, lead cooled fast reactor (LFR) was considered as one of the brightest prospects concept potentially able to excel in four broad areas which are sustainability, economics, safety and reliability, and proliferation resistance and physical protection [3]. Lead-cooled fast reactor features high temperature operation, low pressure, closed fuel cycle and has the ability to operate on a flexible basis between two modes either as breeder to achieve the net creation of fissile fuel or to convert long lived minor actinides and other radioisotopes with to short lived ones as a transmuter. Because of the unique safety benefits of LFR over other fast reactors, it has bright prospects to be under development worldwide. At present, Europe's ELFR lead-cooled system [4], Russia's BREST-OD-300 [5], and the SSTAR system concept designed in the US [6] represents the base of the Generation-IV International Forum (GIF) System Research Plan (SRP) for system concepts of lead-cooled fast reactors [5].

ALFRED (Advanced Lead-cooled Fast Reactor European Demonstrator) is one of the most advanced projects in Europe which will play the role of a demonstrator for the ELFR (European lead fast reactor) concept to prove the safety and reliability in all operating conditions, predicted to start its operation in 2025 [7]. Three of the most important steps of this project are the fuel selection, design and optimization. The goal of this work is to design a 3D-homogeneous and 3D-heterogeneous model for both inner fuel assembly (IA) and outer fuel assembly (OA) to study the neutronic characteristics of the 3D-homogeneous concept relative to the 3D-heterogeneous concept with respect to power distributions, k_∞, reactivity, fuel burn-up and nuclear fuel transmutation. This study represents the first step for further studies to provide alternative approaches of minor actinides management achieving the best refueling scenario in LFR systems.

2. The description of ALFRED core

ALFRED (300 MW_{th}) is a pool-type with small-size. The reactor core is consisting in 171 wrapped hexagonal fuel assemblies (FAs) to form a cylindrical core, together with sixteen positions for twelve control rods (CRs) and four safety rods (SRs), and surrounded by two rows of 108 dummy elements serving as reflector. The 171 Fuel Assemblies (FAs) is divided radially into two zones (inner zone with 57 FAs and outer zone with 114 FAs) with different plutonium enrichments. Each fuel assembly (FA) containing a triangular lattice with 127 fuel pins arranged in it and surrounded by pure lead as coolant. Fig. 1 shows the ALFRED primary system and the layout of the core. The ALFRED core main parameters are presented in Table. 1. The fuel is composed of MOX pellets with theoretical density of 95% and O/M (Oxygen-to-Metal ratio) of 1.97. The U and Pu isotopic vectors have been evaluated considering recovered U and Pu extracted from LWR spent fuel (burnt up to 45 GWd/t, with a 4.5% initial enrichment in ^{235}U) aftera total of 15 years of cooling, 4 of which after reprocessing (so that also the decay of ^{241}Pu into ^{241}Am has been taken into account). See Table. 2 for the fuel isotopic composition of uranium and plutonium. The fuel pin design was considered with inner and outer radius (1, 4.5 mm respectively) and active length of 60 cm [7], [9] as shown in Fig. 2. The design parameters of fuel pin is presented in Table. 3. The considered absorber material for control and safety rods is B_4C (with 90 at.% of ^{10}B and density of 2.2 g/cm^3).

Table 1. ALFRED main parameters [9].

Main parameters	
Thermal power (MW)	300
Fuel assembly concept	Closed hexagonal.
Coolant	Lead
Inlet temperature of coolant (^oC)	400
Outlet temperature of coolant (^oC)	480
Mass flow rate of coolant (kg s^{-1})	26000
Maximum velocity of coolant (m s^{-1})	< 2.0
Cladding	15-15 Ti
Cladding maximum temperature at nominal conditions (^oC)	550

IOP Conf. Series: Materials Science and Engineering 1171 (2021) 012009 doi:10.1088/1757-899X/1171/1/012009

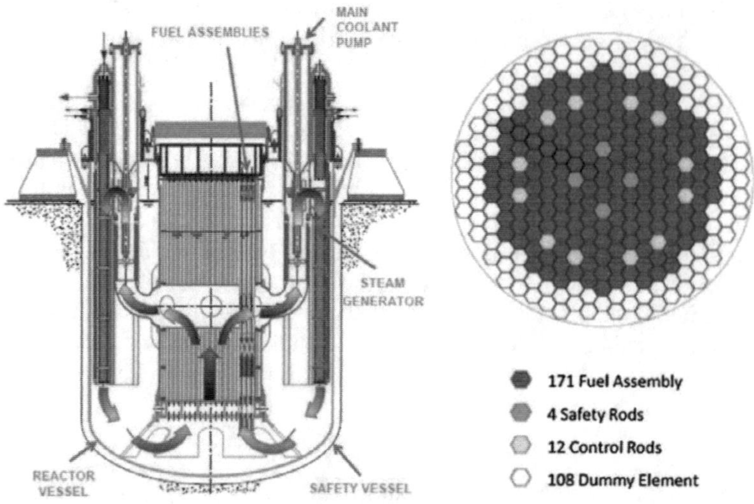

Peak burn-up (MWd/kg)	100

Figure 1. Primary system (left) and core layout (right) of ALFRED [8].

Table 2. The isotopic vectors of uranium and plutonium [11].

Plutonium isotopes	Fraction (wt.%)	Uranium isotopes	Fraction (wt.%)
^{238}Pu	2.332	^{234}U	0.003
^{239}Pu	56.873	^{235}U	0.404
^{240}Pu	26.997	^{236}U	0.010
^{241}Pu	6.105	^{238}U	99.583
^{242}Pu	7.693		

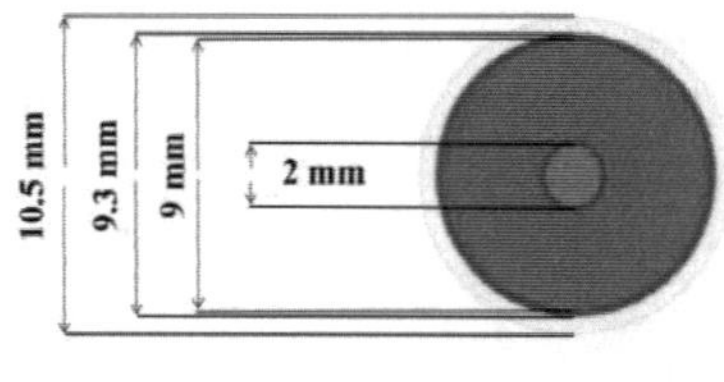

Figure 2. Fuel pin cross section of ALFRED.

3. Modeling and simulation

Two 3D models (3D-heterogeneous and 3D-homogeneous) for both inner fuel assembly (IA) and outer fuel assembly (OA) based on the same design limits and technological constraints have been designed using MCNPX [10] to compare power peaking factor, reactivity parameters, fuel burn-up and masses of fissile inventory. Reflected boundary conditions are assumed for both models (3D-heterogeneous and 3D-homogeneous) to consider interaction of neutrons with other neighboring assemblies in radial direction while the top and bottom boundaries in axial direction are assumed to be free surfaces. MCNPX (Monte-Carlo N-Particle Transport Code) is a general purpose for simulating the interaction of nearly all particles (neutrons, photons and electrons) at nearly all energies with the substance of the system using Monte-Carlo methods. It has the ability to predict the isotopic transmutation of fuel as a function of irradiation time.

Table 3. Design parameters of ALFRED fuel pin [8].

Design Parameters of fuel pin	
Type of fuel	MOX
(Pu+Am)/(Pu+U+Am) enrichment of inner zone (at. %)	21.7
(Pu+Am)/(Pu+U+Am) enrichment of outer zone (at. %)	27.8
Fuel density (%) of theoretical density	95
O/M	1.97
Fill gas	He
Length of upper plenum (mm)	120
Length of Active zone (mm)	600
Length of lower plenum (mm)	550
Outer diameter of cladding (mm)	10.5
Inner diameter of cladding (mm)	9.3
Outer diameter of fuel pellet (mm)	9.0
Inner diameter of fuel pellet (mm)	2.0
Pin pitch (mm)	13.86

3.1. Heterogeneous model of fuel assembly

The 3D-heterogeneous model of fuel assembly is consisting of a triangular lattice with 127 fuel pins arranged in it and surrounded by pure lead as coolant. Then, to enclose the 127 fuel pins and form the bundle, a T91 wrapping with suitable rigidity was added. The fuel pin's design includes a reflector zone that extends 1.0 cm below and above the active length of the fuel (Active zone), as well as an empty plenum zone that extends 12 cm above the higher reflector and 55 cm below the lower reflector providing a confining of the fission the thermal expansion of reflector materials and the fuel as well as the fission gases, after that the reflector extends again for 1.0 cm below the lower plenum and for 6.0 cm above the upper plenum. During the earlier stages of this technology development, Ti-15-15 austenitic stainless steel was proposed as a cladding material in fast reactors. [12], so it has been chosen for cladding the fuel pin. See Fig. 3(a) and Fig. 4(a) for axial and radial cross section respectively of the 3D-heterogeneous model of the ALFRED fuel assembly.

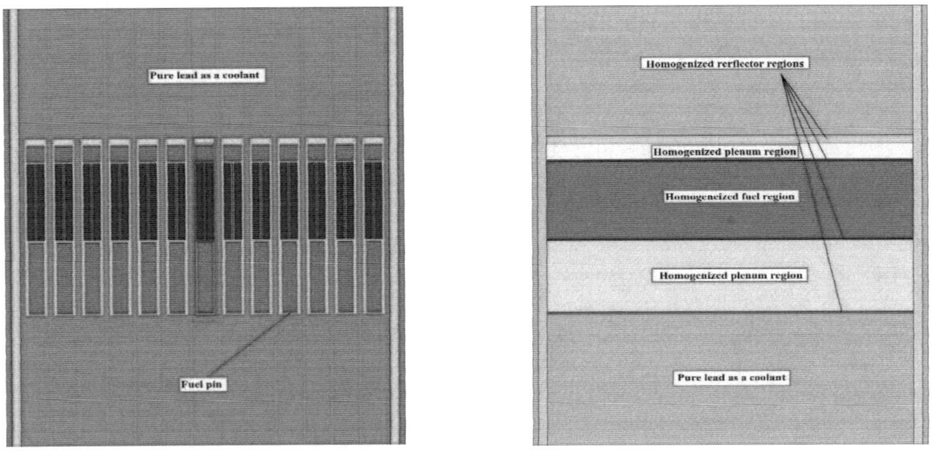

(a) (b)

Figure 3. Axial cross section of ALFRED FA (a) Heterogeneous model (b) Homogeneous model.

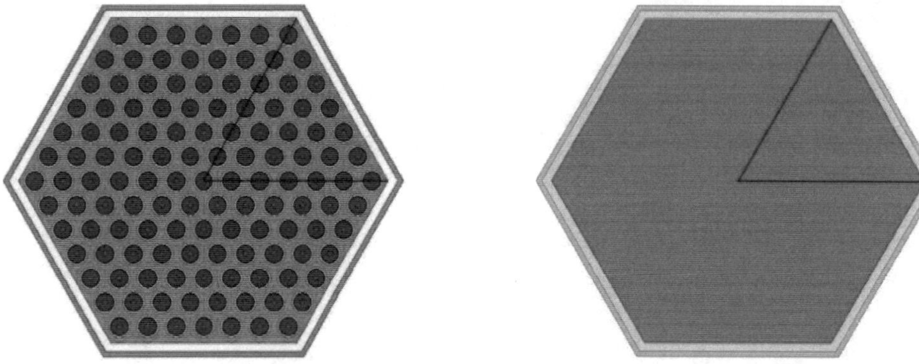

(a) (b)

Figure 4. Radial cross section of ALFRED FA (a) Heterogeneous model (b) Homogeneous model.

3.2. Homogeneous model of fuel assembly

Fig. 3(b) and Fig. 4(b) show axial and radial cross section respectively of the 3D-homogeneous model of the ALFRED fuel assembly. As shown, all the active zone materials within the wrapped hexagonal T91, including the exterior fuel pins with lead coolant were homogenized into one material. In simulations, the active fuel zone was divided into 12 axial meshes of equal length (5 cm each) to analyze the influence of the homogenization on the axial power distribution. Both inner and outer fuel assemblies has been homogenized in order to investigate the influence of fuel homogeneity on neutronic characteristics and fuel performance. This, in turn, helps to provide confidence in fuel burn-up results estimated using 3D-heterogeneous models.

4. Results and discussion

From a neutronic standpoint, this analysis has been characterized by using ENDF/B-VI.2 cross sections library. It includes nuclides cross sections data such as those of fuel, reflector materials, clad and coolant given at multiple temperatures (e.g. 1200 K, 900 K, 600 K and 293 K). The burn-up of MOX fuel inside the assembly is simulated during a total cycle length of 5 years, which is equivalent to 1825 Effective Full Power Days (EFPD), and the build-up of the masses of U/Pu/MA isotopes have been tracked for each irradiation time interval to compare the results of fuel transmutation within 3D-homogeneous relativeto 3D-heterogeneous model for both inner assembly (IA) and outer assembly (OA). The Dell Precision T5610 with two processors; Intel Xeon processor E5-2600 V2 and ECC RDIMM memory of 32GB and 1866MHz has been used to run MCNPX code with 550 cycles. Every simulation calculation goes with the first 55 cycles skipped before the tally accumulation begins. The total number of neutron source histories used in the calculation are 1000. The run time for burn up of IA lasted for 38 hours for the 3D-homogeneous model and 46 hours for the 3D-heterogeneous model. On the other hand, the run time for burn up of OA lasted for about 36 hours for the 3D-homogeneous model and 44 hours for the 3D-heterogeneous model. The results of the simulation will be discussed as follows: the effects of homogeneity on the power distribution for IA and OA are introduced in Section 4.1; the effects of homogeneity on reactivity parameters and fuel burn-up for both IA and OA are discussed in Section 4.2; comparing of fuel transmutation results between 3D-homogeneous and 3D-heterogeneous models for both IA and OA are presented in Section 4.3.

4.1. Power distribution

For both inner assembly (IA) and outer assembly (OA) the fuel region was divided axially into 12

IOP Conf. Series: Materials Science and Engineering 1171 (2021) 012009 doi:10.1088/1757-899X/1171/1/012009

axial regions of equal volumes and radially into 127 hexagonal of equal volumes. The results of comparison for axial power distribution between 3D-homogeneous and 3D-heterogeneous models for IA are slightly different but achieved the same maximum peak power factor. Fig. 5 shows the axial power peaking factor for 3D-heterogeneous and 3D-homogeneous model of IA and OA. The results of comparison for axial distribution between 3D-homogeneous and 3D-heterogeneous model for OA show a good agreement. It is expected because of the lower initial fissile enrichment of the IA in comparison with OA. On the other hand, Fig. 6 shows that, the 3D-heterogeneous model provides at radial power distribution on both IA and OA when compared with the 3D-homogeneous model.

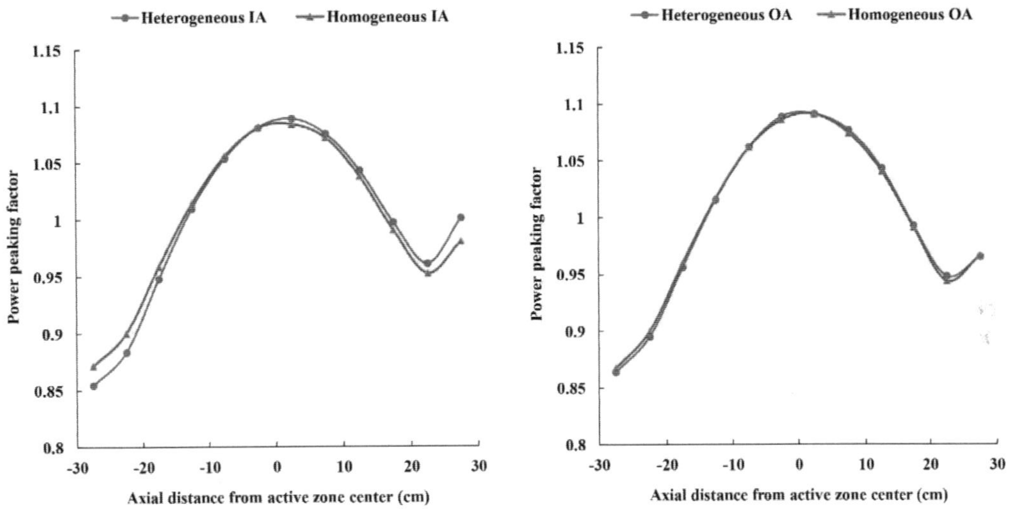

(a) (b)

Figure 5. Axial power distribution for 3D-heterogeneous and 3D-homogeneous models of (a) Inner assembly (IA) (b) Outer assembly (OA).

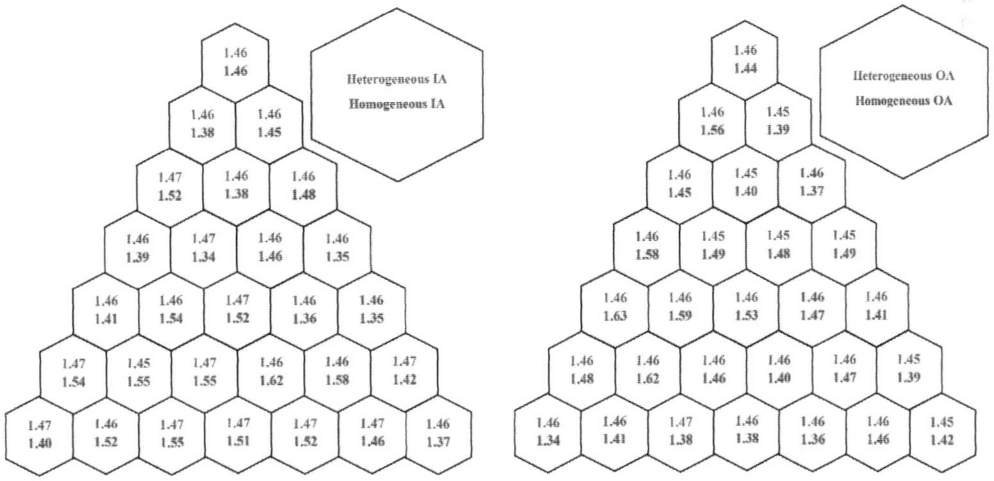

(a) (b)

Figure 6. Radial power peaking factor for 3D-heterogeneous and 3D-homogeneous model of (a) Inner assembly (IA) (b) Outer assembly (OA).

4.2. The effect of homogeneity on fuel burn-up

As shown in Fig. 7 for the k_∞ value as a function of Effective Full Power Days (EFPD) for both the 3D-heterogeneous and 3D-homogeneous models of inner assembly (IA) and outer assembly (OA), the results show that the k_∞ values of 3D-homogeneous assembly is slightly lower than that of 3D-heterogenous for both IA and OA and this is because of the significantly harder spectrum of 3D-heterogeneous model. From a neutronic standpoint, this agreement between the 3D-homogenous and 3D-heterogeneous model can be explained as, for fast neutrons, the length of the elementary lattice pitch is lower than the mean free path of the neutrons.

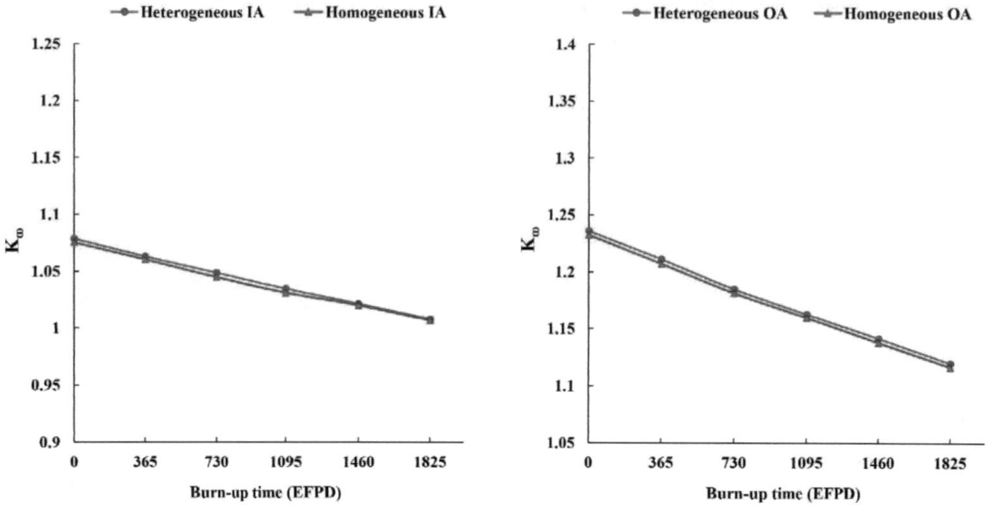

(a) (b)

Figure 7. The effect of homogenization on K_∞ for (a) Inner assembly (IA) (b) Outer assembly (OA).

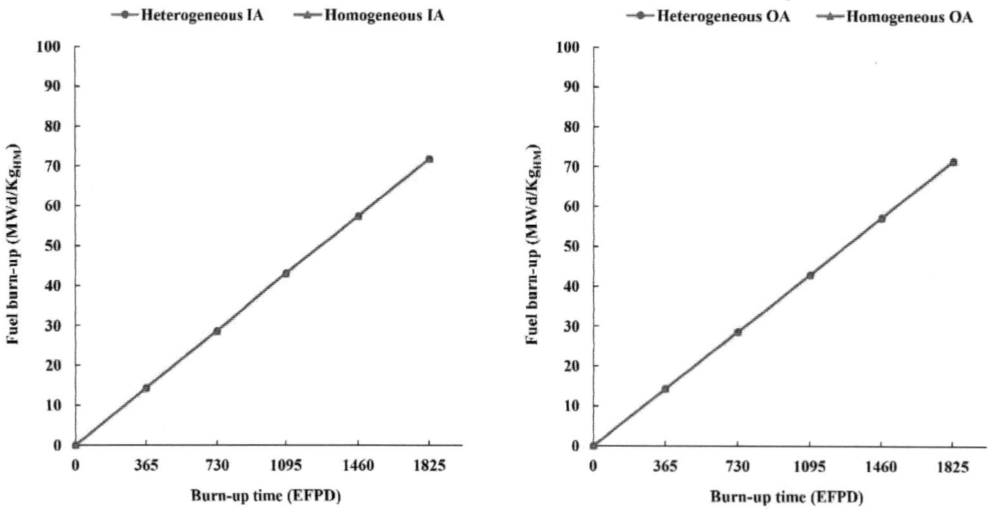

(a) (b)

Figure 8. Peak fuel burn-up for 3D-heterogeneous and 3D-homogeneous models of (a) Inner assembly (IA) (b) Outer assembly (OA).

Fig. 8 shows a good agreement between 3D-heterogeneous and 3D-homogeneous models for fuel burn-up where, the fuel burn-up of the two models has reached 71.8 MWd/kg$_{HM}$ for IA and 71.6 MWd/kg$_{HM}$ for OA at the end of irradiation period (1825 EFPD).

4.3. *The effect of homogeneity on fuel transmutation*

As shown in Fig. 9(a) and Fig. 9(b), the ^{235}U mass inventory of 3D-heterogeneous and 3D-homogeneous model shows good agreement for both inner assembly (IA) and outer assembly (OA). As we can see from Fig. 10(a), for IA, the total Pu mass inventory of 3D-homogeneous model is higher than the 3D-heterogeneous models at the end of irradiation time.

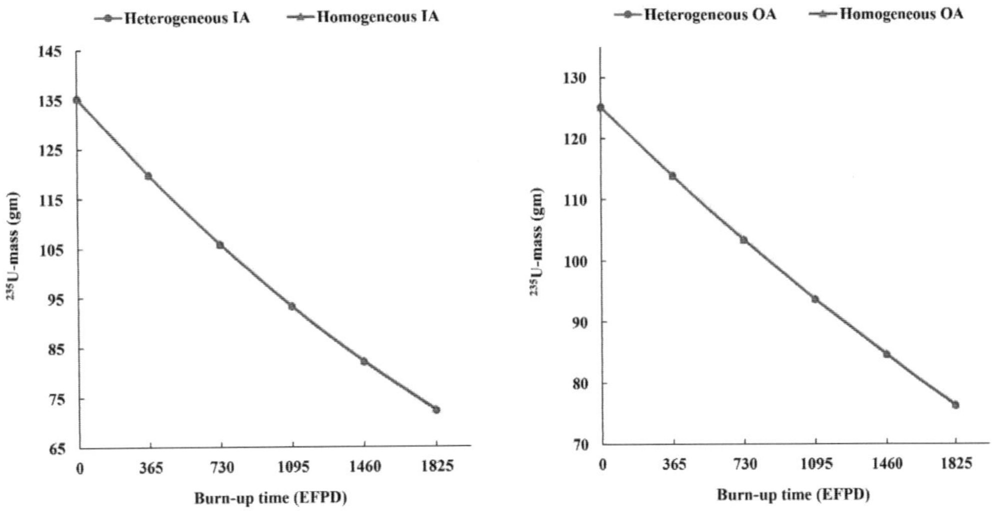

(a) (b)

Figure 9. ^{235}U mass inventory for 3D-heterogeneous and 3D-homogeneous models of (a) Inner assembly (IA) (b) Outer assembly (OA).

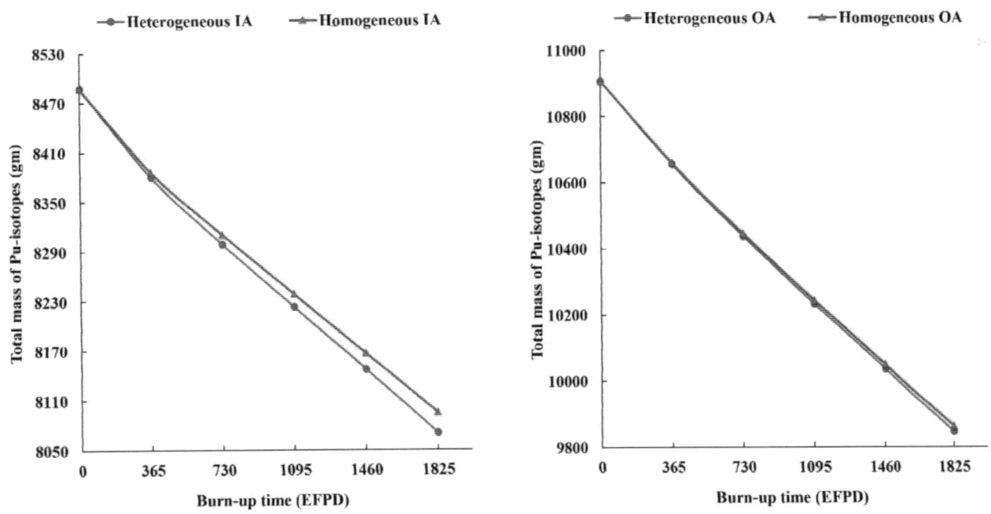

(a) (b)

Figure 10. Total mass inventory of Pu isotopes for 3D-heterogeneous and 3D-homogeneous models of (a) Inner assembly (IA) (b) Outer assembly (OA).

On the other hand, for OA, Fig. 10(b) shows a good agreement between 3D-heterogeneous and

3D-homogeneous model for the total mass inventory of Pu isotopes. This is because the results indicate that, the destruction rate of fertile Pu isotopes (^{239}Pu and ^{241}Pu) of 3D-homogenous model is lower than of 3D-heterogenous model for IA. This indicates that the destruction rate of fertile nuclides (except ^{242}Pu) of 3D-heterogeneous model is higher than of 3D-homogeneous model for IA due to the low enrichment of Pu in IA relative to Pu enrichment in OA. See Fig. 11 and Fig. 12.

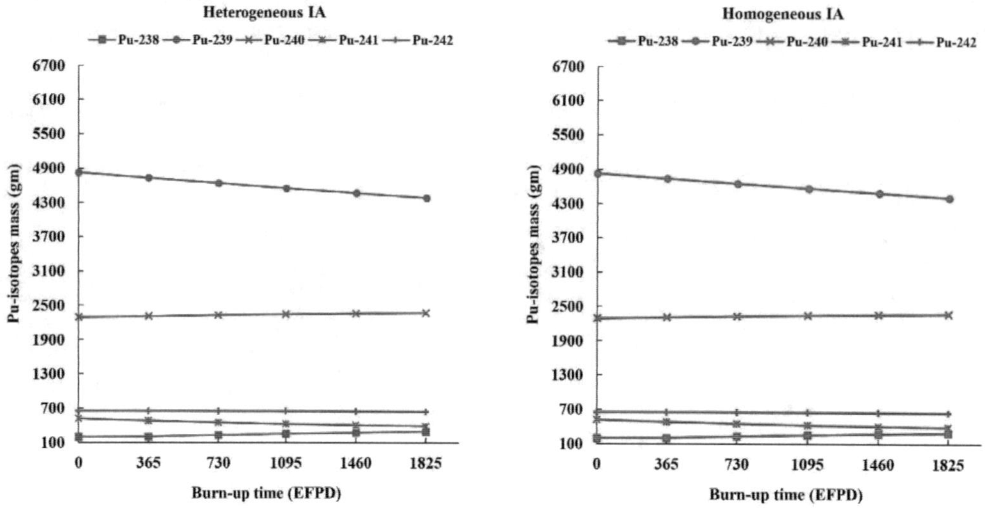

(a) (b)

Figure 11. Mass inventory of Pu isotopes of inner assembly (IA) for (a) 3D-heterogeneous model (b) 3D-homogeneous model.

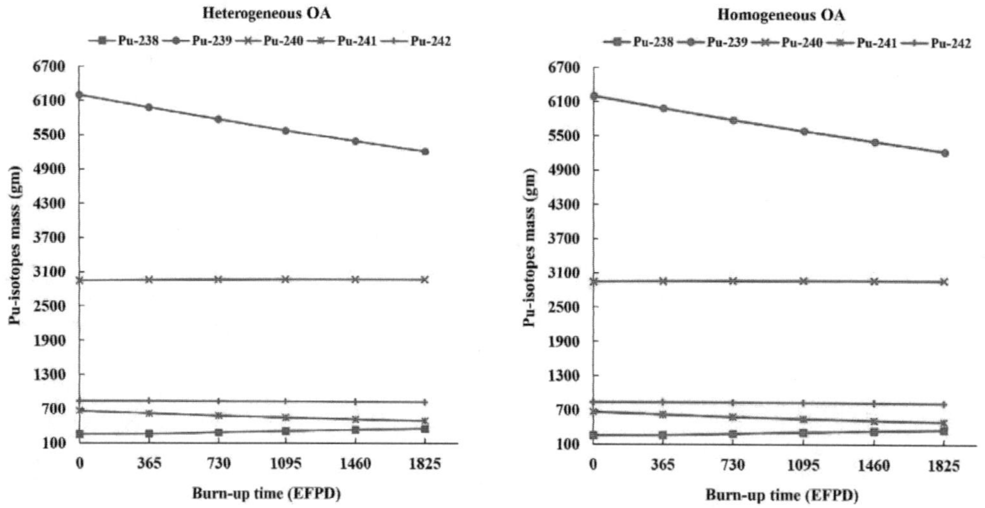

(a) (b)

Figure 12. Mass inventory of Pu isotopes of outer assembly (OA) for (a) 3D-heterogeneous model (b) 3D-homogeneous model.

Fig. 13 shows good agreement between 3D-heterogeneous and 3D-homogeneous model for both IA and OA with irradiation time for ^{241}Am. Also, you can see how Fig. 14 and Fig. 15 show good agreement for the MA-isotopes mass inventory with irradiation time for both IA and OA.

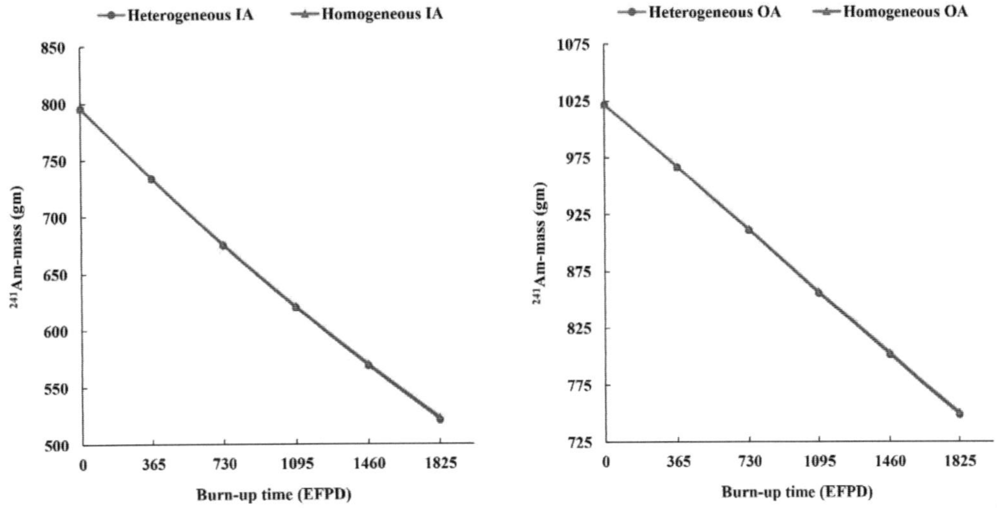

(a) (b)

Figure 13. ^{241}Am mass inventory for 3D-heterogeneous and 3D-homogeneous models of (a) Inner assembly (IA) (b) Outer assembly (OA).

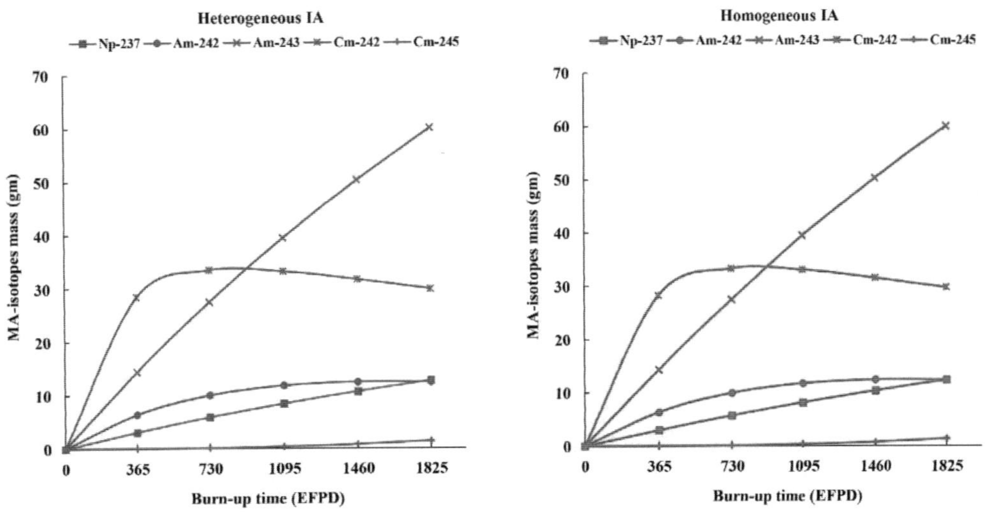

(a) (b)

Figure 14. MA mass inventory of inner assembly (IA) for (a) 3D-heterogeneous model (b) 3D-homogeneous model.

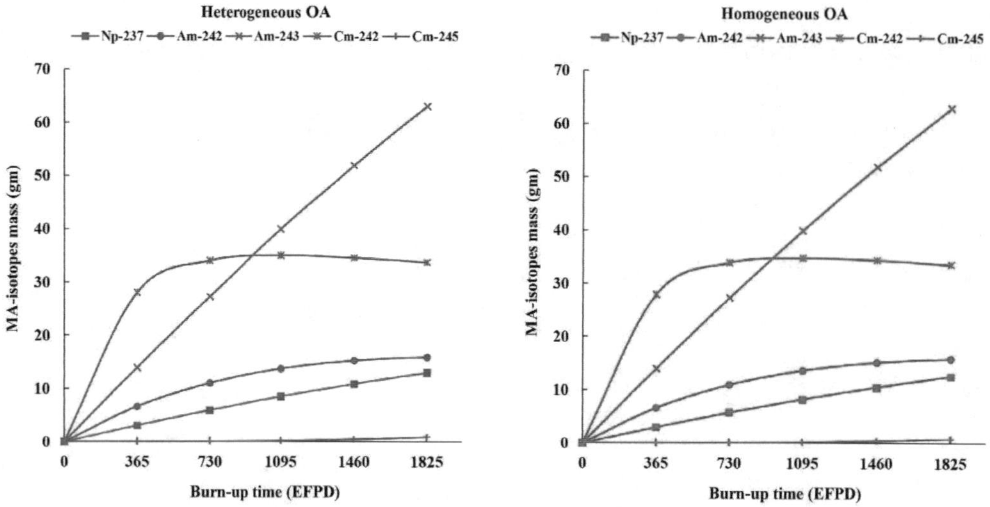

(a) (b)

Figure 15. MA mass inventory of outer assembly (OA) for (a) 3D-heterogeneous model (b) 3D-homogeneous model.

5. Conclusion

Two models (3D-heterogeneous and 3D-homogeneous) for both inner fuel assembly (IA) and outer fuel assembly (OA) have been designed using MCNPX to study homogenization effect on power peaking factor, k_{∞}, fuel burn-up and the fuel isotopic transmutation. Both 3D-homogeneous and 3D-heterogeneous designs were based on the same design assumptions and constraints. Firstly, the 3D-homogeneous model can be employed as a beginning steps for validation and verification calculations rather than 3D-heterogeneous model by observation that it saves about 18% of run time for burn-up simulation for both IA and OA. The neutronic analysis shows close behavior between the two models of axial power distribution for both IA an OA which has a peak of (1.09). On the other hand, the 3D-heterogeneous model provides at radial power distribution on both IA and OA when compared with the 3D-homogeneous. There is no effect of homogenization on the fuel burn-up of for both IA and OA at the end of burn-up time (1825 EFPD). The results of fuel transmutation for IA show that the destruction rate of fertile nuclides (^{235}U, ^{239}Pu, ^{241}Pu) of 3D-homogeneous model is lower than of 3D-heterogenous model. This characteristic would allow the 3D-homogeneous model to have a longer fuel operating cycle than the 3D-heterogeneous model. The results show close behavior for the transmutation rate of MA for 3D-homogeneous model relative to the 3D-heterogeneous model for both IA and OA.

References

[1] DOE, U. S. (2002). A technology roadmap for generation IV nuclear energy systems.
[2] Waltar, A. E., Todd, D. R., & Tsvetkov, P. V. (Eds.). (2011). *Fast spectrum reactors.* DOI 10.1007/978-1-4419-9572-8. Springer Science & Business Media.
[3] Zohuri, B. (2020). *Nuclear Reactor Technology Development and Utilization* (pp. 61-120). Woodhead Publishing.
[4] Alemberti, A., Carlsson, J., Malambu, E., Orden, A., Cinotti, L., Struwe, D., ... & Monti, S. (2011). *Journal of nuclear science and technology*, *48*(4), 479-482.
[5] Zrodnikov, A. V., Toshinsky, G. I., Komlev, O. G., Stepanov, V. S., & Klimov, N. N. (2011). *Journal of Nuclear Materials*, *415*(3), 237-244.

[6] Smith, C. F., Halsey, W. G., Brown, N. W., Sienicki, J. J., Moisseytsev, A., & Wade, D. C. (2008). *Journal of Nuclear Materials*, *376*(3), 255-259.
[7] Grasso, G., Petrovich, C., Mattioli, D., Artioli, C., Sciora, P., Gugiu, D., ... & Mikityuk, K. (2014). *Nuclear Engineering and Design*, *278*, 287-301.
[8] Luzzi, L., Cammi, A., Di Marcello, V., Lorenzi, S., Pizzocri, D., & Van Uffelen, P. (2014). *Nuclear Engineering and Design*, *277*, 173-187.
[9] Grasso, G., Petrovich, C., Mikityuk, K., Mattioli, D., Manni, F., & Gugiu, D. (2013). *Proceedings of Fast Reactors and Related Fuel Cycles, Vol. 2: Safe Technologies and Sustainable Scenarios (FR13), Paris.*
[10] Pelowitz, D. B., Durkee, J. W., Elson, J. S., Fensin, M. L., Hendricks, J. S., James, M. R., ... & Verbeke, J. M. (2011). Los Alamos National Lab.(LANL), Los Alamos, NM (United states).
[11] Sobolev, V., Malambu, E., & Abderrahim, H. A. (2009). *Journal of Nuclear Materials*, *385*(2), 392-399.
[12] Bortot, S., Cammi, A., Lorenzi, S., Ponciroli, R., Della Bona, A., & Juarez, N. B. (2013). *Nuclear Engineering and Design*, *265*, 1238-1245.

IOP Conf. Series: Materials Science and Engineering 1171 (2021) 012010 doi:10.1088/1757-899X/1171/1/012010

Shielding parameters for cobalt free steel alloys

M M Mourad[1], H A saudi[2], M M Eissa[3]and M Y Hassaan[4]and A Abdel-Latif M[5]
[1]Ministry of electricity and renewable energy
[2] Physics Department, Faculty of Science, Al-Azhar University (Girls Branch), Nasr City 11884, Cairo, Egypt
[3] Department of Steel and Ferroalloys, CMRDI, Helwan, Egypt
[4] Physics Department, Faculty of Science, Al-Azhar University (Boys Branch), Nasr City 11884, Cairo, Egypt
[5] Department of Physics, Faculty of Engineering, Fayoum University, Fayoum 63511, Egypt

Abstract. Steel alloys are widely used for radiation shielding in nuclear applications since Cobalt is an expensive element, this leads to the steels is may be expensive, so it is preventing wider application and selection. So the important direction of this research is preparing cobalt-free maraging stainless steel as shielding to reduce the production cost. Therefore, seven different free-cobalt steel alloys were prepared by using an electro slag re-melting technique. Steel compound ratios were calculated by using the software WinXCOM program for Monte Carlo simulation, at energies of photon 662, 1173 and 1332 keV. The attenuation properties of these alloys were studied. Furthermore, the total of removal macroscopic cross-section, transmission number and mean free path were determined using Geant4 code for fast neutrons radiation shielding. Therefore, shielding parameter variations are applied to the steel alloys to investigate the superior shielding properties to gamma rays than other materials.

Keywords. Free cobalt stainless steel, Coefficients of mass attenuation, Fast neutrons, Removal cross-section

1. Introduction

Gamma and neutron radiation are employed in many applications like high-energy physics, material analysis, and nuclear plants [1-24]. According to this purpose; using the radiations increases every day in technology, and the need for electrical energy increases, leading to more and more new nuclear power plants [2–3]. According to this purpose; different types of stainless steel were developed, containing of heavy elements such as iron, barium, or titanium in their structure [4–5]. This steel has well mechanical properties at high temperatures, so it may be used for radiation protection in many applications in nuclear energy reactors. The steel is that the protected material within the transfer of heat system within the core of the reactor and also the neutron modulation system. The coefficient of attenuation is a very dominant parameter which distinguishes between the spread and gamma rays and penetration of neutrons with the matter, and depends on alloys chemical compositions with falling energy [22]. The scattered and absorbed radiation is relevant to the electron effective density and the matter density. Using 137Cs and 60Co sources for gamma rays to calculate linear mass attenuation coefficients at energies of photon from 662 keV to 1.33 MeV [6-20]. Therefore, the aim of the present work is producing cobalt-free alloys composite capable of use as a nuclear radiation shield, which is better than other alloys. From the result, the total removal cross-section, transmission number and mean free path values of studied steel alloys are a good type of stainless steel for shielding specially (0.032 % Ni, 0.009 % Mo, 0.087 % Cr, 0.948 % Mn, 0.002 % Ti, 0.005 % V, 97.01 % Fe, 0.3 % C, 0.035 % P, 0.012 % S with minor elements to reach 100% of the composition) with density 8.28 g/cm^3 compared with the different steel alloys used in a convenient material shielding in the field of nuclear.

2. Methods and materials

2.1 Preparation sampling of the Material Shielding

The preparation samples of free cobalt stainless-steel done by induction the furnace of pilot plant electro-slag re-melting technique (ESRT). We used Chromium and titanium instead cobalt element to outdo the issue of austenitic retained. The percentage content of nickel reduced to 14 % by weight. Materials mixed for 15min and became a homogeneous mixture; it heated at 350°C and pelletized pressure at 600 MPa. The formed samples pellet annealed at 820 °C for one hour and repeating this process for two hours at 1300 °C [7-12], and then annealed studied samples of steel alloys were hardened by the faster cooling process.

2.2 Mass Attenuation Coefficient

By using the method of gamma transmission, we used a detector (HPGe) with relative efficiency 25% relative to the NaI(Tl) detector for measuring the coefficient of mass attenuation for gamma radiation with a collimated narrow beam. The sample located among the standard source of gamma radiation and detector. Repeat this process once with the sample and once without it for 5 minutes. The samples irradiated by emitted photons from (137 Cs and 60 Co) sources radioactive from 662 keV to 1.33 MeV [6-14]. Using the Lambert-Beer law for gamma-ray photons mono-energetic parallel beam was attenuated with the matter [15].

$$I(x) = I_0 \exp(-\mu x) \quad (1)$$

Where I_0 is the density of initial photon, I(x) is the collimated beam which photons are penetrated thickness x and μ is the total linear attenuation coefficient. So the equation may be written for the collimated beam by:

$$\frac{\mu}{\rho} = \frac{Ln\left(\frac{I_0}{I}\right)}{\rho x} \quad (2)$$

The mass attenuation coefficient μ/ρ for mixtures of elements can be calculated by eq.[16-17]

$$\frac{\mu}{\rho} = \Sigma iwi \left(\frac{\mu}{\rho}\right) \quad (3)$$

Where ρ is partial density, $\left(\frac{\mu}{\rho}\right)_i$ is the mass attenuation of the ith constituent and w^i is the weight fraction of ith constituent. Also, the half-value layer (HVL) can be calculated:

$$HVL = \frac{Ln(2)}{\mu} \quad (4)$$

2.3 Monte Carlo simulation codes Geant4

Geant4 is a software tools for the simulation of the particles movement through matter and it used by a huge number of projects and experiments in several application fields such high energy physics, astrophysics and space science, medical physics, and radiation protection. Its operation and modeling capabilities continue to be extensive, while its performance enhanced. An overview of recent developments in diverse areas of the toolkit is presented. These include performance optimization for complex setups; improvements for the propagation in fields; new options for event biasing; and additions and improvements in geometry, physics processes and interactive capabilities [19]. The simulation Geant4 code is depend on the targeted material type, the particle geometry and various energies of photon. The simulation experimental setup explains what is happening between targeted material and radiation. It used in high energy physics, nuclear applications and in the, materials research, medical physic and particle accelerator [8-13].

2.4 Theoretical basis

2.4.1 The effective removal cross-section of Fast neutrons (Σ_R) cm^{-1}

The effective removal cross-section is called also attenuation coefficient of neutron (Σ_R) for homogenous mixtures or compounds, and it can be determined by using Kaplan eq. [18].

$$\sum_R/\rho = \frac{Ln(\frac{I_0}{I})}{\rho x} = \sum_i W_i \left(\frac{\Sigma R}{\rho}\right)_i \quad (5)$$

$$\sum_R = \frac{Ln(\frac{I_0}{I})}{\rho x} = \sum_i \rho_i \left(\frac{\Sigma R}{\rho}\right)_i \quad (6)$$

Where W_i is the percentage weight, x is the alloy thickness, ρ_i is the partial density and $(\Sigma_R/\rho)_i$ is the fast neutron mass attenuation coefficient of the ith constituent.

2.4.2 The Total macroscopic cross-section ($\sum_t$) cm^{-1}

The total of macroscopic cross section is a very significant parameter for neutron shielding particles, and this theoretically values was calculated for energy 4.5 MeV fast neutrons by using simulation of Monte Carlo Geant 4 code. The results are showed in Table 4. The cross-section used for explain the reactions possibility among neutrons and target material in particle physics and nuclear physics [9-12], how a particle or a neutron can be interacting with the material target categorized by the cross-section. Neutron reaction probability with the light ions (alpha particles) or nucleus by a microscopic cross-section (σ) and the interaction probability with the heavier materials are calculated by the macroscopic cross-section. Neutrons can interact with the target material such as absorption, scattering, fission, capture, etc.. So the total macroscopic cross-section ($\sum_t$) can be determined by equation [9-10].

$$\sum t = N\sigma \quad (7)$$

Where σ is the microscopic cross-section unit with m^{-1} and N is the atomic density of target material which given by:

$$N=(\rho/A).NA \quad (8)$$

As ρ: density target material, NA : Avogadro number.

$$\sum \text{Total} = \sum \text{scattering} + \sum \text{absorption} + \sum \text{capture} + \sum \text{fission} \ldots \quad (9)$$

Materials that have large total cross-section are excellent moderators for neutrons that means the high possibility interaction with the target materials [9-10].

The mean free path λ is the path traveled by a neutron among two collisions or interactions with the target material and the collision probability P(x) at a distance x taken by a neutron in the material target can be determined [13-15].

$$p(x)dx = \sum_t \exp^{-\Sigma_t} xdx \quad (10)$$

(x): The distance of neutrons this can be taken in the material without any interaction. The mean free path (λ) may be calculated by using:

$$\lambda = \int_0^\infty x\, p(x)dx = \frac{1}{\Sigma_t} \quad (11)$$

$\sum_t$: The interaction possibility of neutron per unit length [11], transmission number of neutron that passing and incoming is explained by transmission 100,000 fast neutrons which sent to each material and the numbers of neutrons passing through the materials can be calculated by (Geant 4 code) Monte Carlo simulation.

3. Result and discussion

In this research, the seven new different samples of high-alloyed steel were produced, the chemical compositions of them showed in table1.

Table 1: Chemical compositions and densities of produced alloys.

sample	Elements (wt. %)										Density (g/cm^3)
	Ni	Mo	Cr	Mn	Ti	V	Fe	C	P	S	
s_1	0.032	0.009	0.087	0.948	0.002	0.005	97.01	0.3	0.035	0.012	8.28
s_2	12.655	5.565	5.15	0.258	0.0045	0.033	75.03	0.05	0.014	0.009	7.93
s_3	12.61	5.76	5.3	0.275	0.0075	0.037	74.62	0.04	0.013	0.005	7.91
s_4	13.48	3.915	9.04	0.2865	0.1165	0.0735	71.42	0.02	0.015	0.005	7.8
s_5	12.225	5.305	5.74	0.31	0.414	0.052	74.26	0.03	0.01	0.005	8.12
s_6	10.46	4.585	17.03	0.491	0.4265	0.1205	63.97	0.04	0.011	0.011	7.67
s_7	11.825	3.595	13.24	0.371	0.1075	0.097	67.55	0.02	0.015	0.011	7.77

The gamma rays attenuation coefficient of the prepared different steel alloys observed with energy range 622 - 1333 keV. The intensities of gamma rays that transmitted through the steel alloys samples was determined and given as samples thickness function. The curves of attenuation were used to find the linear attenuation coefficients average values and hence the coefficients of experimental mass attenuation $(\mu/\rho)_{Exp.}$ also found. These values of $(\mu/\rho)_{Exp.}$ tabulated in Table2 and it compared with the theoretical coefficients of mass attenuation $(\mu/\rho)_{Theo.}$ which completed by using WinXCOM software computer [24]. It observed that the coefficient of mass attenuation has nearly the same behavior for all investigated steel alloys. Also the coefficients of mass attenuation have nearly the same values for all the prepared samples. This behavior was mainly due to the balance in the high value of different Z-elements (Cr, Ni, Mn and Fe) of the steel grades.

Table 2: The mass attenuation $(\mu/\rho)_{Exp}$ and $(\mu/\rho)_{Theo}$ for the studied alloy with different thickness (t).

Sample	Energy line (keV)	Mass attenuation $(\mu/\rho)_{Exp}$ (cm^2/gm)			Mass attenuation $(\mu/\rho)_{Theo}$ (cm^2/gm)
		t = 0.43 cm	t = 0.868 cm	t = 1.308 cm	
S_1	662	0.0734±0.009	0.0736±0.006	0.0733±0.009	0.0735
	1173	0.0551±0.001	0.0555±0.005	0.0552±0.001	0.0553
	1333	0.0518±0.001	0.0519±0.002	0.054±0.004	0.0518
S_2	662	0.0736±0.008	0.0734±0.009	0.0730±0.009	0.0738
	1173	0.0555±0.005	0.0552±0.001	0.0549±0.005	0.0547
	1333	0.0520±0.002	0.0517±0.001	0.0515±0.001	0.0512
S_3	662	0.0722±0.007	0.0737±0.007	0.0734±0.009	0.0737
	1173	0.0542±0.004	0.0555±0.005	0.0552±0.001	0.0553
	1333	0.0508±0.001	0.0519±0.001	0.0516±0.001	0.0518
S_4	662	0.0736±0.006	0.0737±0.003	0.0830±0.01	0.0737
	1173	0.0558±0.008	0.0561±0.006	0.0558±0.008	0.0553
	1333	0.0520±0.002	0.0518±0.002	0.0517±0.001	0.0519
S_5	662	0.0737±0.007	0.0739±0.004	0.0735±0.009	0.0737
	1173	0.0554±0.004	0.0555±0.005	0.0553±0.004	0.0553
	1333	0.0518±0.001	0.0519±0.002	0.0517±0.002	0.0518
S_6	662	0.0736±0.006	0.0747±0.003	0.0744±0.008	0.0735
	1173	0.0553±0.003	0.0562±0.006	0.0559±0.006	0.0552
	1333	0.0517±0.001	0.0525±0.002	0.0523±0.002	0.0518
S_7	662	0.0735±0.009	0.0735±0.009	0.0733±0.009	0.0736
	1173	0.0555±0.005	0.0555±0.005	0.0551±0.005	0.0553
	1333	0.0519±0.002	0.0517±0.001	0.0516±0.001	0.0518

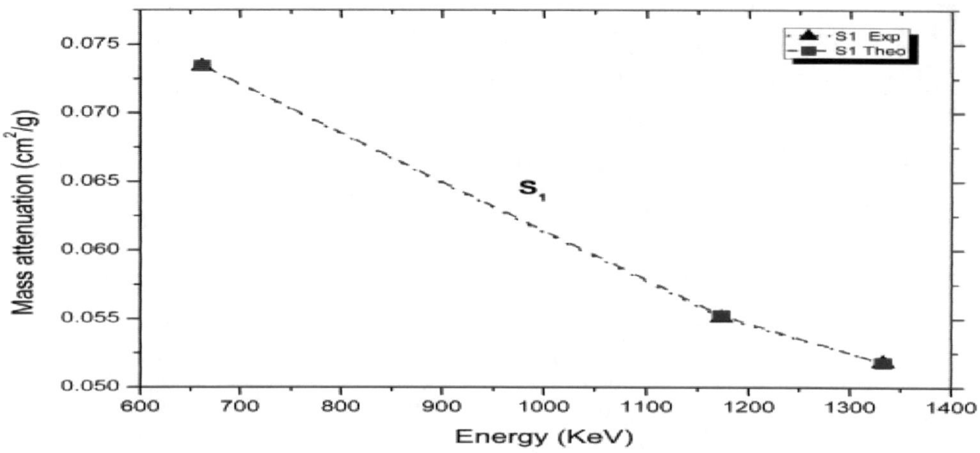

Figure1. The relation among mass attenuation (exp. and theo) with different energies for sample 1.

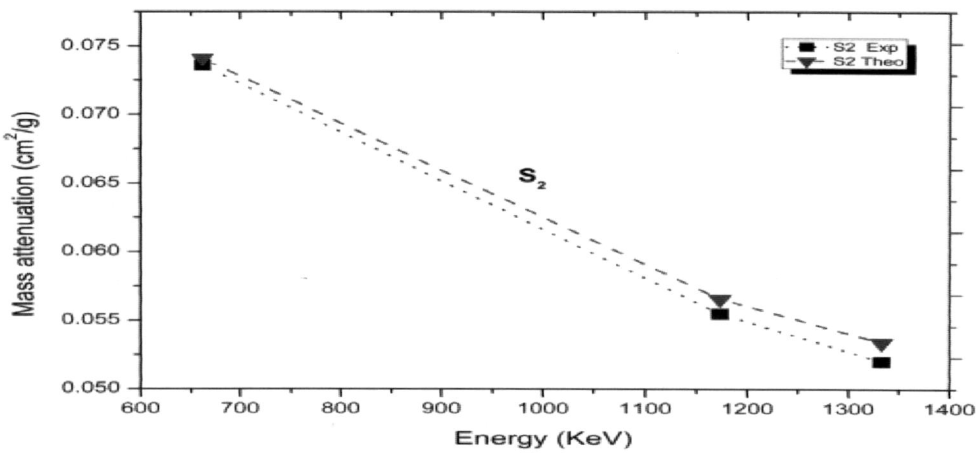

Figure2. The relation among mass attenuation (exp. and theo) with different energies for sample 2.

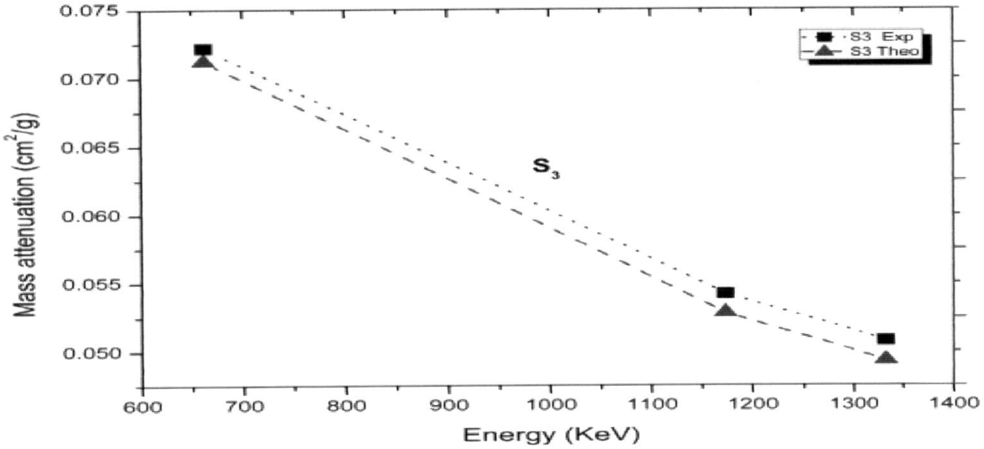

Figure3.The relation among mass attenuation (exp. and theo) with different energies for sample 3.

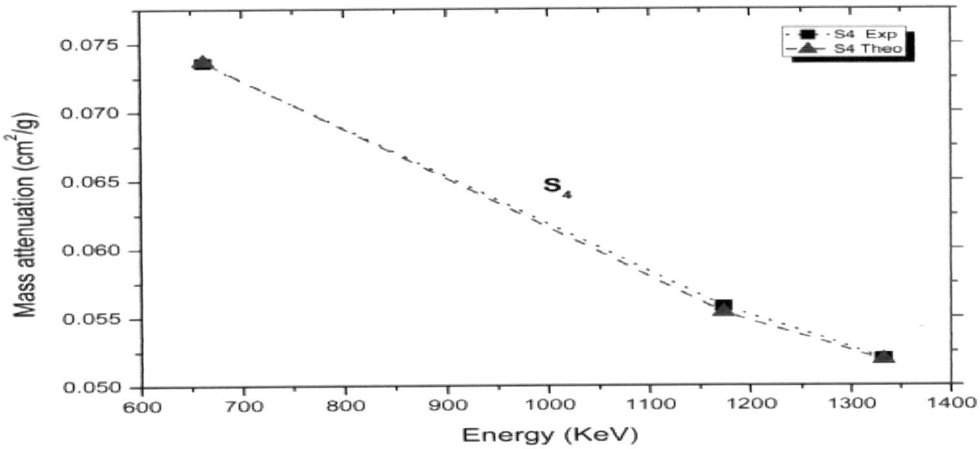

Figure 4.The relation among mass attenuation (exp. and theo) with different energies for sample 4.

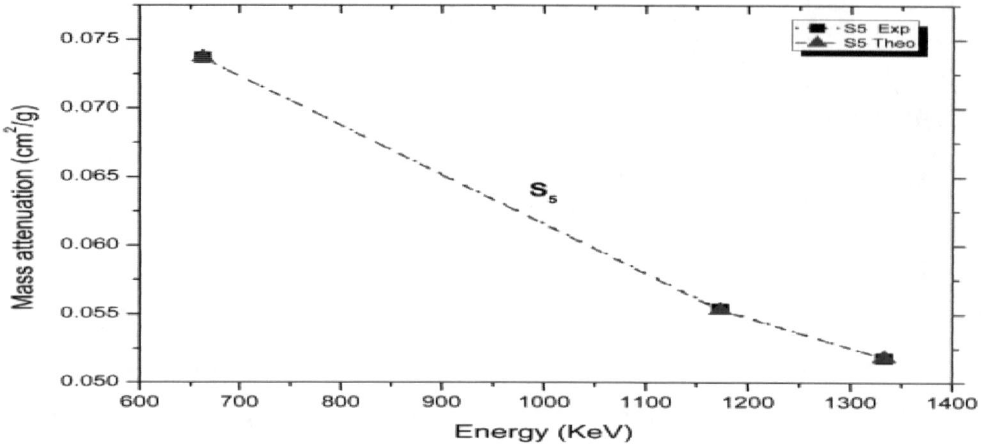

Figure 5. The relation among mass attenuation (exp. and theo) with different energies for sample 5.

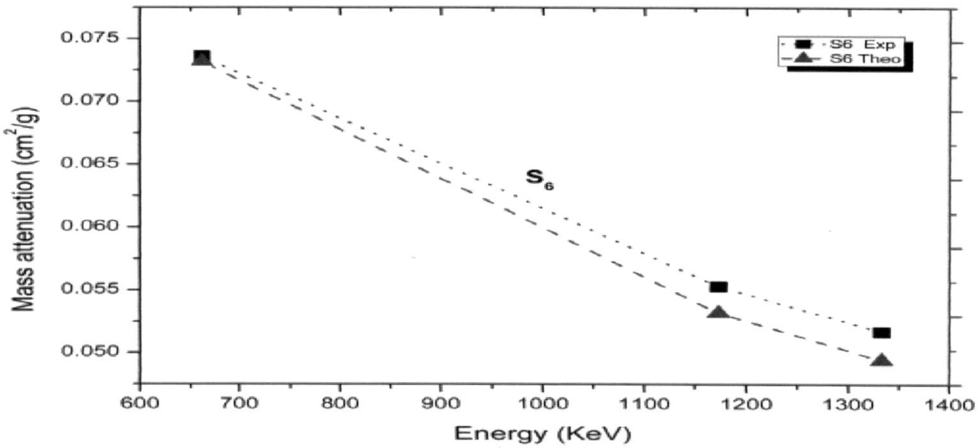

Figure 6.The relation among mass attenuation (exp. and theo) with different energies for sample 6.

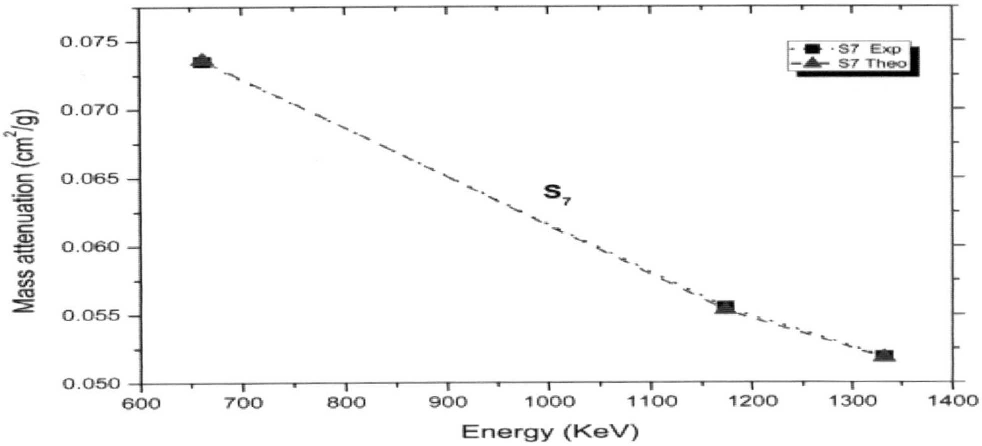

Figure7. The relation among mass attenuation (exp. and theo) with different energies for sample 7.

The observed variations in the coefficients of mass attenuation as shown in the previous figures (1: 7) of different seven studied samples and can be explained by photon energy and Z-dependency of an interaction elements cross-section. There were two different regions observed in the behavior of the coefficients of mass attenuation. First region is a sharp decrease of coefficients of mass attenuation in the energy range up to 622 keV because of the dominant reaction among gamma rays and the investigated alloys grades is the photoelectric effect. Second region is a slight decrease in the coefficients of mass attenuation from 911.2 keV up to 1333 keV, which may be considered this process of Compton scattering [19]. Additionally, it is excellent agreement among the experimental results of coefficients of mass attenuation and corresponding theoretical data which determined by the WinXCOM software. Half value layer is the required thickness to half the intensity of coming radiation. It observed in table 3 the sample number 1 has the smallest "HVL" among samples that examined. Samples that have smallest value of " HVL" means that have more ability for absorption the radiation.

Table 3: Half value layer at different thickness of studied samples.

Sample	Energy line (keV)	HVL		
		t = 0.43 cm	t = 0.868 cm	t = 1.308 cm
S_1	662	1.140	1.14	1.71
	1173	1.518	1.51	2.27
	1333	1.616	1.61	2.43
S_2	662	1.188	1.19	1.79
	1173	1.576	1.58	2.39
	1333	1.682	1.69	2.55
S_3	662	1.213	1.19	1.79
	1173	1.616	1.58	2.38
	1333	1.725	1.69	2.55
S_4	662	1.208	1.21	1.82
	1173	1.593	1.58	2.39
	1333	1.709	1.71	2.58
S_5	662	1.158	1.16	1.74
	1173	1.541	1.54	2.32
	1333	1.647	1.64	2.48

S_6	662	1.227	1.21	1.82
	1173	1.634	1.61	2.42
	1333	1.746	1.72	2.59
S_7	662	1.213	1.21	1.83
	1173	1.608	1.61	2.42
	1333	1.719	1.72	2.60

Also a total macroscopic cross-section ($\sum t$)cm^{-1} for fast neutrons, mean free pass (λ) was determined by simulation of Monte Carlo Geant 4 code as shown in table 4.

Table 4: Total macroscopic cross-section ($\sum t$) and mean free path (λ).

Sample no.	Total macroscopic cross-section ($\sum t$) cm^{-1}	Mean free path (λ) cm
S_1	0.1871	5.344
S_2	0.1697	5.893
S_3	0.1677	5.963
S_4	0.1638	6.105
S_5	0.1721	5.809
S_6	0.1608	6.217
S_7	0.1632	6.129

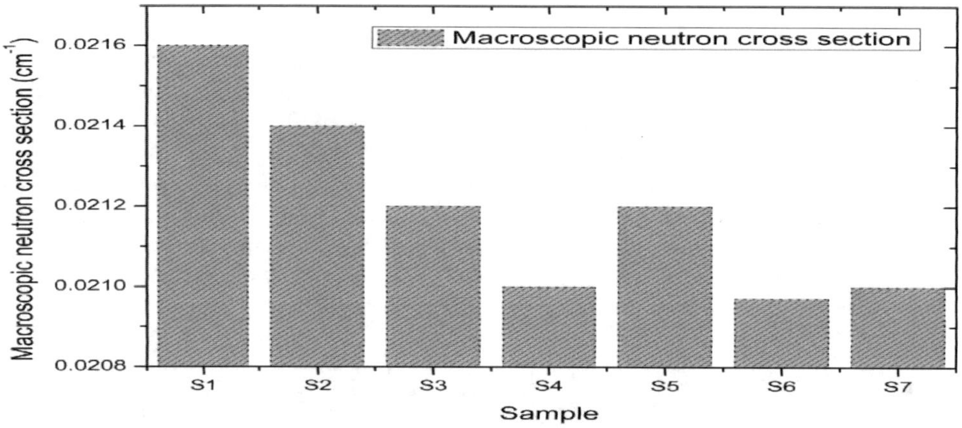

Figure 8. Total macroscopic cross section data of the seven studied alloys at different energies.

The macroscopic cross section of studied seven samples at different energies shown in Fig. 8 and it clear that the macroscopic cross section range of seven samples is about range 0.1608 to 0.1871 and it was considered a good type of stainless steel alloys for shielding of fast neutrons radiations. According to different thickness (0.43-0.868-1.308 cm) the transmission amount dose which passes calculated by Simulation of Monte Carlo Geant 4 code and as tabulated in table5.

Table 5: Transmission number with different thickness.

	Transmission number for samples						
Thickness(cm)	S_1	S_2	S_3	S_4	S_5	S_6	S_7
0.43	92269	92963	93043	93199	92865	93318	93224
0.868	85008	86303	86454	86747	86121	86970	86794
1.308	78289	80094	80305	80715	79838	81028	80781

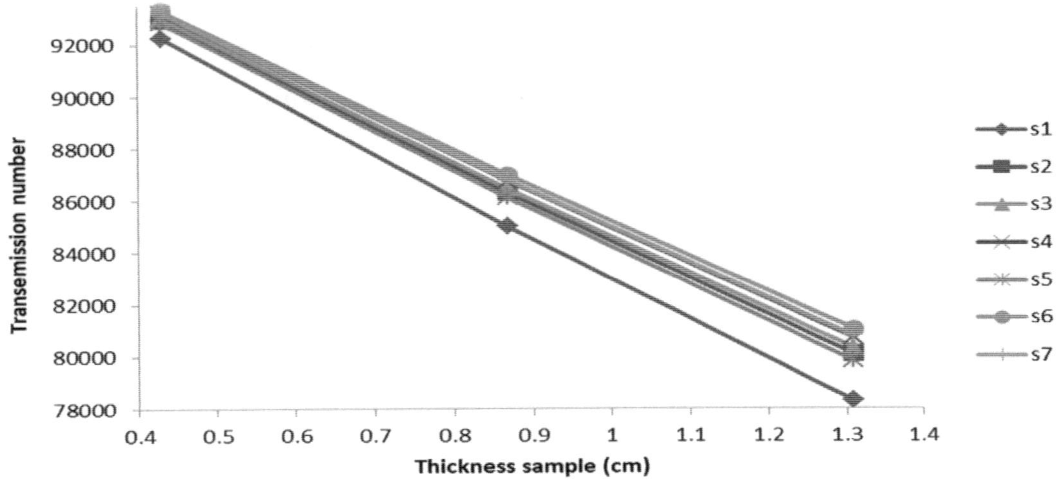

Figure 9. Transmission number data of the seven studied alloys at different thickness.

According to thickness of samples; the neutron transmission changed. 100,000 fast neutrons with energy 4.5 MeV was sent for each neutrons number and thickness was determined which passing through the material. It observed in Fig.9 that as all sample's thickness increases and the neutrons number which passed was obviously reduced. Clearly, the new high alloys stell are perfect as material shielding ability radiation for fast neutron.

4. Conclusion

In this work, the Mass attenuation coefficient (cm2/g) was calculated at three different energies of gamma-ray for new steel samples, and half values layer (HVL) was calculated for the seven samples. Total macroscopic cross-section, transmission number and mean free path values determined for 4.5 MeV particles fast neutron. Neutron transmission numbers at 0.43 cm material thickness were calculated and obtained that there is decreasing in the neutrons number for all new alloys. From the result; it observed that both of the total cross-section, transmission number and mean free path of the stainless-steel highly alloyed samples were considered a good type of stainless steel alloys and promising for use as shielding alloys for gamma-ray and fast neutrons radiations especially sample 1 (0.032 % Ni, 0.009 % Mo, 0.087 % Cr, 0.948 % Mn, 0.002 % Ti, 0.005 % V, 97.01 % Fe, 0.3 % C, 0.035 % P, 0.012 % S).

5. References

[1] Nuclear Regulatory Commission. (1997). *United States Nuclear Regulatory Commission information digest: 1997 edition. **Volume 9*** **(No NUREG--1350-VOL. 9).** Nuclear Regulatory Commission..

[2] Singh, V P Badiger, N M **(2014)** *Annals of Nuclear Energy*, ***64*****, 301-310.**

[3] Betterodge, W & Heslop J (**1974**).

[4] Fattah-Alhosseini, A, Shirsalimi, F, Yousefi, M, & Abedi, A **(2014)** *J Mater Environ. Sci*, *5*(6), **1847-1853.**

[5] Yonezawa T (2012).

[6] Kwon, M E, Bae, Y S, Cho, S Y, Choe, W H, Hong, B G, Hwang, Y S, ... & Yu, I K (2009) *Nucl ar Engineering and Technology*, ***41*****(4), 455-476.**

[7] Aygün, B, Şakar, E, Korkut, T, Sayyed, M I, & Karabulut, A (2019). *Radiochimica Acta*, ***107*****(4), 359-367.**

[8] Agostinelli, S, Allison, J, Amako, K A, Apostolakis, J, Araujo, H, Arce, P, ... & Geant4 Collaboration (2003) *Nuclear instruments and methods in physics research section A: Accelerators, Spectrometers, Detectors and Associated Equipment*, ***506*****(3), 250-303..**

[9] Mughabghab, S F (2012*). Academic press.*

[10] McLane, V., Dunford, C. L., & Rose, P. F. **(1988).**
[11] Rinard, P. (1991). *Passive nondestructive assay of nuclear materials*, **(375-377)**.
[12] Phase, B V (**2006**. *Washington, DC: The British Institute of Radiology.*
[13] Shore, R. E., Beck, H L, Boice, J D, Caffrey, E. A Davis, S, Grogan, H A, ... & Dauer L T (2018).*Journal of radiological protection*, *38*(3), **1217**.
[14] Mukherjee, S, & Jamnapara, N I (2015, September). In *Proc Indian Natn Sci Acad* **(Vol. 81, No. 4, pp. 827-839).**
[15] Wood, J **(2013)** *Elsevier.*
[16] Akkurt, I, & El-Khayatt, A M (2013) *Journal of Radioanalytical and Nuclear Chemistry*, ***295*(1), 633-638.**
[17] Kurudirek, M, Özdemir, Y, & El-Khayatt, A M (2011). *Radiation Physics and Chemistry*, ***80*(8), 855-862.**
[18] Gong, J, Si, W, & Guan, Z (2001) *Journal of non-crystalline solids*, ***282*(2-3), 325-328**.
[19] Ott, K O, & Bezella, W A (1983). *Amer Nuclear Society.*
[20] Collaboration, G E A N T, & Agostinelli, S (2003).*Nucl. Instrum. Meth.* ***A*, *506*(25), 0.**
[21] Al-Hadeethi, Y, & Sayyed, M I (2020) *Progress in Nuclear Energy*, ***129*, 103511.**
[22] Aşkın, A (2020) *Ceramics International*, *46*(9), **14090-14096**.
[23] Boukhris, I, Kebaili, I, Al-Buriahi, M. S, & Sayyed, M I (2021) *Journal of Non-Crystalline Solids*, ***551*, 120430**.
[24] Kassab, M M, El-Kameesy, S U, & Eissa, M M (2015) *Journal of Modern Physics*, ***6*(11), 1526**

Promoting the physical and the mechanical properties of concrete using nano and waste materials

Ahmed A. AbdelGhafaar[a], Ahmed S. Faried[a], M. Mokhtar[b], M. Morsy[b], Khaled M. Osman[a]
[a] Faculty of Engineering, Fayoum University, Fayoum, Egypt
[b] Building Physics and Environment Institute, Housing & Building National Research Center (HBRC), Dokki, 12311 Giza, Egypt

Abstract: the industry of producing Portland cement concrete (PCC) depends on Portland cement (PC) as a major element. One of the main usages of the PCC is producing concrete members. The use of waste materials in concrete manufacture could be useful to improve the properties of concrete and reduce production costs by replacing the Cementous material with waste materials. Researchers had invested efforts in investigating the effect of using nanomaterials as an add-on to the concrete mixture on the mechanical properties of concrete. As a result, it was concluded that nanomaterials have a considerable impact on promoting the compressive and flexural strength of the PCC. This research illustrates the effect of utilizing nanomaterials, as an add-on to the concrete mixture, combined with using waste materials, as a replacement to PC, to improve the mechanical properties of plain concrete with different types of materials in different dosages. The experimental part has been carried out by using waste materials and nano-based material. The used waste materials were ceramic waste (CW), and marble powder (MP). The nano-based materials that were used were Carbon Nanotubes (CNTs) and Graphene oxide (GO). The results were compared in terms of compressive strength and flexure strength to the control concrete mix. These tests aim to investigate the impact of the modified mixtures on the mechanical properties after elapsing 7 and 28 days to obtain the maximum partial replacement percentage of cement with waste and nano-based materials. It could be concluded that using CW and MP combined with either CNTs or GO could be effectively utilized as a replacement of cement in PCC.

KEYWORDS: Waste Material, Cement, Marble powder, Ceramic Waste, graphene oxide, Carbon Nanotubes.

1. INTRODUCTION

Portland cement concrete (PCC), as a construction material, is widely used to resist loads with sufficient compressive strength. Adding waste and nano-based material could lead to obvious enhancement in mechanical properties. The necessitates achieving obvious enhancement in the mechanical properties cannot always be achieved using conventional mixes. Another challenge is to reduce the amount of the Cementitious material in order to reduce the production cost of PCC without reducing the concrete performance, especially for big projects that require a large amount of PC [1–3]

Efforts have been made to search for admixtures to enhance the performance of the concrete. According to experimental trials, waste materials could be used in the concrete to achieve better performance and better mechanical properties. Using Waste materials in concrete could lead to improve the properties of concrete, and reduce the cost of production as a replacement of PC with waste materials. If locally available waste materials are used, it may be possible to reduce the cost of various construction works. Using waste material as a replacement for cement in concrete has a positive side as it reduces energy consumption; moreover, it has a noticeable positive impact on the environment. Furthermore, it also has an effect on the economic side by decreasing concrete costs as the cost of Cementous material approximately equals more than 45% of PCC cost. [4].

The waste of ceramic is about 15%-30% of the material generated from the total production. Out of total waste, a large amount disposed as waste. Ceramic waste (CW) can be categorized into two types, waste earthenware and fractured during the industrialization process. Ceramic waste can be considered a harmless waste that possesses pozzolanic characteristics. [5,6].

On the other hand, the marble and granite industry are common materials that are widely used in the construction industry. Some studies estimate that approximately 20–25% of the total production of marble and granite are produced in the powder form. Dumping these stones' wastes is a major environmental issue because of its highly alkaline nature, and its manufacturing techniques, which cause a health hazard to the surroundings. [7–9]

Nowadays, the application of nanotechnology has shown a remarkable result in solving numerous engineering-related problems. Nanomaterials could enhance material properties, for example, thermal, electrical, and mechanical. When Compared to the micro/macro-sized particles, the aspect ratio (surface area to volume ratio) of the nanoparticle is higher; thus, Nanomaterials could enhance the cementitious material properties.[10–15]

Graphene is recognized as a material that has an outstanding material property, and it is known to have a high possibility of being used in many industries, for example, electronics, optics, sensors and polymer composites. Graphene is much more robust than steel about 200 times. Graphene Oxide (GO) is derived from graphene.
Graphene oxide has a noticeable effect on cement's microstructure and mechanical properties. GO has unique physical properties that could make it an effective material to be used in concrete mixture as a replacement for cement.[16]

CNTs is getting increasing scientific and commercial attention because of its particular chemical, and physical characteristics that qualify it to be an appropriate option for that could be used in many fields, for instance, electronic materials, energy, chemistry, and In Civil Engineering applications due to its unique mechanical properties, low specific weight, and high resistance to corrosion. In addition to their unique strength, which is approximately equal 100 times that of the tensile strength of steel at one-sixth of the weight, CNT has shown an outstanding array of other properties. As a result, CNT is efficiently used in various research works.[17]

Many research efforts have been made recently in order to investigate the effect of implementing nanotechnology in recent researches, and it has been reported that nanomaterials have a notable effect in enhancing the strength and durability of PCC. Furthermore, many researches have been conducted on using waste material to improve the mechanical properties and to reduce the cost of using the Cementous material. This paper investigates the possibility of using both waste material (ceramic waste and marble powder) and nanomaterials (graphene oxide and nanocarbon tubes) as a blend in different dosages in order to be used as a replacement for cement for waste material and as an add-on for nanomaterial.

Accordingly, this proposed work aims to evaluate the impact of utilizing CW and MP in varying percentages as a partial replacement of PC and utilizing Go or NCT as an add-on to the modified concrete mix. The experimental work has been carried out in three phases. In phase 1, CW and MP were used separately with 5%, 10%, 15%, and 20% by weight of cement while in phase 2, the best ratio of CW and MP was used separately in conjunction with a constant ratio equal .05% of nanocarbon tube by weight of cement. In phase 3, the best ratio of CW and MP was used separately in conjunction with a constant ratio equal .05% of Graphene oxide by weight of cement.

2. PROPERTIES OF MATERIALS

The materials which have been used in this study are coarse aggregate, fine aggregate and additives such as superplasticizer, with (Naphthalene Formaldehyde Sulphonate) as a chemical base, Ceramic Waste (CW), Marble Powder (MP), Carbon Nanotubes (CNTs) and Graphene oxide (GO). The properties of these materials were verified and tabulated.
In order to design the concrete mix of PCC, American code for concrete mix, ACI-211-11, was used. The CW and MP were obtained from an industrial sector in Egypt. Specimens of wastes were gathered manually and stored according to the standard specification.

2.1 Cement

The cement used in this research is manufactured by a local cement factory in Egypt and known commercially as "Beni Suef cement." 52.5 grades (ASTM Type II) of the ordinary Portland cement have been utilized. Table 1 represents the physical properties, and Table 2 represents the chemical composition of the used sample of cement.

Table 1. Physical Properties of Cement

Property	Value
Specific Gravity	3.15
Initial Setting Time (min)	45
Final Setting Time (min)	600
Autoclave expansion	≤ 0.16 (%)
Fineness	≥ 2900 (cm^2/g)

Table 2. Chemical Compositions of Cement

Material	Weight (%)
SiO_2	26.02
Fe_2O_3	0.10
CaO	67.95
Al_2O_3	5.86
MgO	0.07

2.2 Coarse Aggregate (Crushed Basalt Rock)

In this research, the coarse aggregate used is fractions of maximum size of 10 mm obtained from Suez crushed Dolomit rock, Egypt. The physical properties of coarse aggregate were determined following the test methods specified in ASTM as shown in Table 3.

Table 3. Physical Properties of Coarse Aggregates

Property	Value
Specific Gravity	2.22
Bulk Density (kg/m^3)	1527
Water absorption (%)	0.15

2.3 Fine Aggregate

The fine aggregate used in the experimental works is a fine aggregate river sand for Beni Suef, Egypt. The fractions of the fine aggregate are of a max size 2.7. The physical properties of fine aggregate were determined ASTM as shown in Table 4.

Table 4. Physical Properties of Fine Aggregate

Property	Value
Specific Gravity	2.86
Bulk Density (kg/m^3)	1642
Water absorption (%)	2

2.4 Waste Materials

2.4.1 Ceramic Waste (CW)

CW used in this experimental study were obtained from The manufacturing waste of the "Ceramica Cleopatra" factory. To get the collected samples ready to be used as a powder, Broken pieces were crushed and then ground with an air jet mill. Finally, the resulted powder was sieved through 200 a 75-lm (200 mesh) sieve. Table 5 shows the chemical compositions of CW.

Table 5. Chemical Compositions of CW

Material	Weight (%)
SiO_2	65.36
Fe_2O_3	4.52
K_2O	1.72
Na_2O	1.99
CaO	1.56
Al_2O_3	18.42
TiO_2	0.74
MnO	0.07
P_2O_5	0.03
SO_3	0.06
MgO	0.77
LOI	1.23

2.4.2 Marble Powder (MP)

Table 6. Chemical Compositions of MP

Material	Weight (%)
SiO_2	1.12
Al_2O_3	0.73
Fe2O3	0.05
CaO	83.22
MgO	0.52
SO_3	0.56
K_2O	0.09
Na_2O	1.12
Free CaO	0.15
L.O.I	2.5
Ins.R	0.89

2.4.3 Carbon Nanotubes (CNTs)

The CNTs have been synthesized by using chemical vapor deposition (CVD) method which is developed at HBRC, Egypt. The CNTs had inner diameters of 9 nm and functionalized with COOH functional groups via strong acid treatment. More details regarding synthesize process can be found in our previous work [18,19].

2.4.4 Graphene oxide (GO)

Graphene oxide has been synthesized from graphite flaks by modified Hummer method. All details related to the synthesizing process were reported in our previous work [18,20][21]

3. EXPERIMENTAL INVESTIGATION

The objective of this research is to achieve the maximum percentage of the partial replacement of cement using different dosages of CW, MP, combined with a constant ratio of CNTs and GO as an add-on to the modified mix. Furthermore, we investigate the impact of modified mixtures on the mechanical properties after 7 days and 28 days.

Subsequently, concrete mixes using various mix proportions for ceramic powder and marble powder were implemented. Then, CW and MP were used separately in conjunction with a constant ratio equal .05% of either CNTs or GO. Different PCC concrete mixes were optimized using waste materials and nono materials in three phases and were used in this research.

3.1 Mix Types and Proportion

The experimental work, using CW, MB, CNTs, and GO conducted in three phases as follows:

- *Phase 1,*
  - A- CW was used with 5 wt %, 10 wt%, 15 wt% and 20wt% of cement as a replacement of PC
  - B- MP was used with 5wt %, 10 wt %, 15 wt % and 20 wt % of cement as a replacement of PC
- *Phase 2,*

A- CW with 20 wt % , of cement as a replacement of PC, was used in conjunction with a constant ratio equal .05 wt % CNTs of cement as an additive.
B- MP with 20 wt %, of cement as a replacement of PC, was used in conjunction with a constant ratio equal .05 wt % CNTs of cement as an additive.

- *Phase 3,*
  A- CW with 20 wt %, of cement as a replacement of PC, was used in conjunction with a constant ratio equal .05 wt % GO of cement as an additive.
  B- MP with 20 wt %, of cement as a replacement of PC, was used in conjunction with a constant ratio equal .05 wt % GO of cement as an additive.

In addition to the mentioned three phases, a control mix was molded with zero percentage of CW, MB, CNTs, and GO. It is noteworthy that phase 3 was designed based on the results of phases 1 and 2, as described later in this research. Table 7 summarizes the different phases mix.

Table 7. Mix Types and Proportion

Phase	Mix Type	Cement, %	CW, %	MP, %	CNTs, %	GO, %
Control Mix		100	0	0	0	0
	1-A-1	95	5			
	1-A-2	90	10	0		
	1-A-3	85	15			
Phase (1)	1-A-4	80	20		0	0
	1-B-1	95		5		
	1-B-2	90	0	10		
	1-B-3	85		15		
	1-B-4	80		20		
Control Mix		100	0	0	0	0
Phase (2)	2-A-4	80.00	20	0	0.05	0
	2-B-4	80.00	0	20		
Control Mix		100	0	0	0	
Phase (3)	3-A-4	80.00	20	0	0	0.05
	3-B-4	80.00	0	20		

3.2 Mixture Design

ACI-211-11 was used as a reference to design the mixture. The values of the binder content and water to cement ratio (W/C ratio) were chosen to be constant and equal 350 kg/m^3 and 0.5, respectively.

Standard slump cone was used to measure the fresh concrete workability according to the slump test methodology as per ASTM C 143. In the beginning, the coarse and the fine aggregate were mixed, then cement, waste material, nanomaterial, superplasticizer, and water were added.

In order to confirm that the nanomaterials (CNTs and GO) have been entirely dissolved in the water, nanomaterials were dispersed in water for 10 min using ultrasonic process model VCX 500.

3.3 Compressive and flexure and Strength

Steel cubic molds with dimensions (10X10X10cm) were used for casting the test samples that were used for compressive strength test. Wooden forms with dimensions (10X10X50cm) were used as molds for casting the test sample to be used for investigating the flexure strength. The specimens were well compacted using a vibrating table. The specimens were left in the mold for 24 h then they were removed from the molds. A water tank at a temperature of 21° C was used for the curing process of the specimens. The specimens were left in the tank until the testing time. Casting, compaction, and curing were performed as per ASTM C 192.

The compressive strength was obtained at 7 and 28 days and the flexure strength at 28 days. For all mixes, the compressive strength was specified by obtaining the average of three cubes.

3.4 Instrumental Analyses

- The surface morphology of casted concrete samples was investigated by the scanning electron microscope (SEM). The SEM microphotographs were obtained with Inspect S (FEI Company, Holland) equipped with an energy dispersive X-ray analyzer (EDXA).
- The microstructure of the utilized nano materials (CNTs and GO) was examined by HRTEM model FEI Philips Tecnai G2 S-Twin operated at 200 keV.
- The Raman spectroscopy as a powerful diagnostic technique for the identification of various types of carbon nano based materials was utilized. The spectra is collected by (SENTERRA–Bruker, Germany) instrument with laser source (Nd:YAG) at wavelength of 532 nm.

4. RESULTS AND DISCUSSIONS

4.1. Structural analysi

The HRTEM photo of CNTs represented in fig. 1a exhibited a tubular structure with internal hollow cavity. The photo confirmed that the CNTs are of multi walled carbon nanotubes (MWCNTs). Fig. 1b depict he microstructure of GO as examined by HRTEM. The examined GO has thin wrinkled and corrugated morphology. This morphology is due to the oxidation process and transformation of planar sp^2 carbon sheets to sp^3-hybridized carbon sheet.

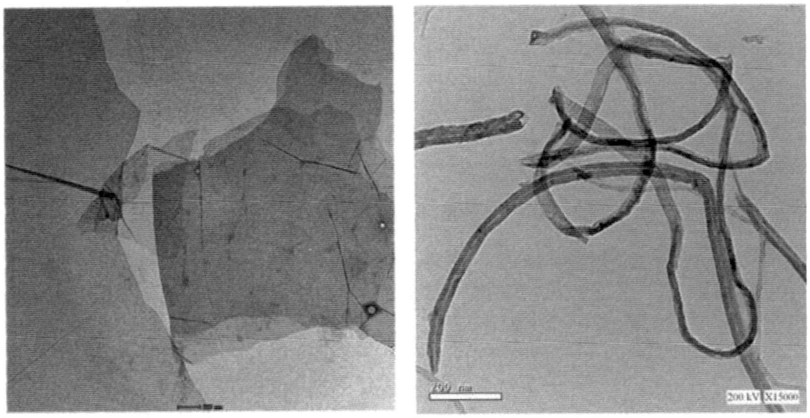

Fig. 1: HRTEM of a) graphene oxide, and b) CNTs

4.2. Phase 1

Group 1A

The curves of compressive strength of 1A group at 7 and 28 days are shown in figure 2, while the reduction and increment in compressive strength as compared to control mix is represented in fig. 3. It can be seen that from fig. 3 at the age of 7 days, the compressive strength of the group 1A (containing CW) increases up to 10 % replacement of CW, then at 15 % CW the compressive strength decrases, with further increases in the amout of CW up to 20% the strength regained approximately the same value as 10% CW replacement. While after 28 days of hydration, the compressive strength was increased with increasing the amount of CW until 10% replacement. With further increasing in the amount of CW the compressive strength starts to be deteririorate. When compared to control mix an enhancement up to 3% in compressinve strength has been recogonized for 10% CW relacement after 28 days of hydration as represented in fig. 3.

The increament in compressive strength up to 10%CW replacemnt could be explained based on the good pozzolanic nature of CW as it contains a high amount of silica. When OPC is mixed with water it forms C-S-H gels and liberates $Ca(OH)_2$. The liberated $Ca(OH)_2$ reacts with SiO_2 constituent of CW to form secondary C-S-H gels. This in turn will resulted in consuming harmful $Ca(OH)_2$, therby reducing the prosity and hence increasing the compressive strength [22]. The decrement in compressive strength for 1-A-3 and 1-A-4 could be attributed to the presence of high amount of silica that can not find a sufficient amount of Ca $(OH)_2$ to react with, in addtion to othe agglomeration of the CW resulted from its high repacement ratio in the concrete matrix [23].

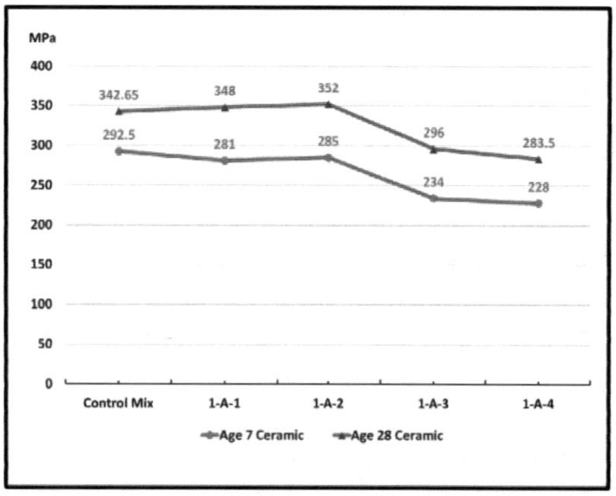

Figure 2: Compressive Strength for Different Mixes of Phase 1-A

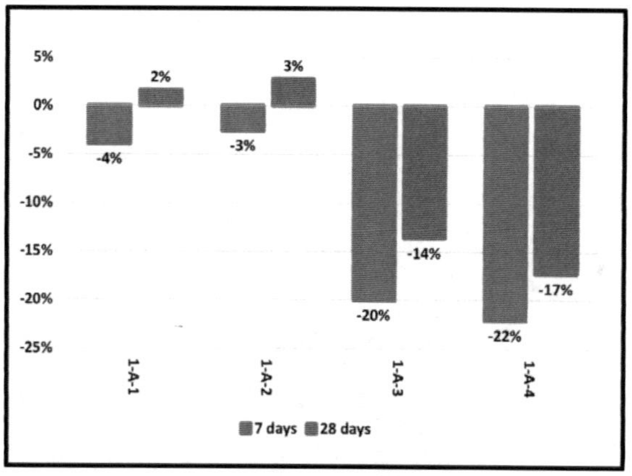

Figure 3: Reduction/Increase of Compressive Strength for Different Mixes of Phase1-A

The felxural strength as represented in fig.4 increases up to 10% CW replacement of cement, with further increases in the amount of CW, the felxural strength decreases. This behaviour is in a good agreement with ompressive strength data.

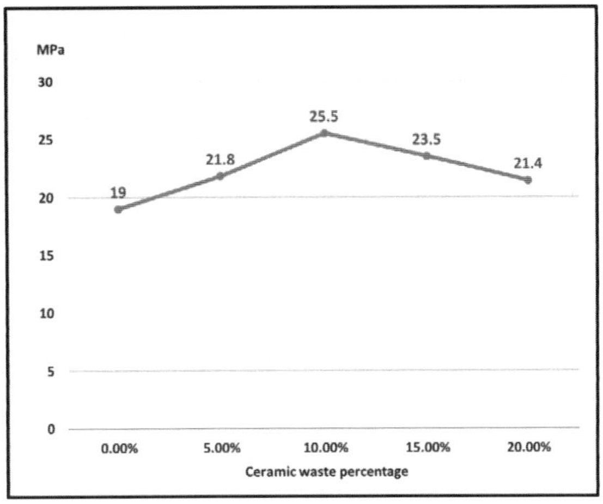

Figure 4: Flexure Strength for Different Mixes of Phase 1-A

Fig. 5: SEM image of control mix

The SEM image of control mix represented in fig. 5, show to some extent the presence of micropors, cracks, separation between the aggregat and the other concrete components. All of these features led to lack of homoginity and compaction of the structure. The deterioration in the mechanical stength, at 20%CW can also confirmed using SEM as represented in fig. 5. The SEM image exhibited a loose structure with agglomerated CW particles which have irregular and angular morphology that analogous to the morphology of cement grain. It was also observed that the size of voids have been increased in addition to the separation between matrix and CW became more prominent.

Group 1B

The compressive strength curves of the group 1B that contains MP after 7 and 28 days of hydration is depectited in fig. 6. It was observed that after 7 days of hydration the compresive strength starts to decreases with 5% MP replacement, then at 10% MP the compressive strength has the same value of control mix. with further repacement in MP, the compresevie strength deteriorated again. Generally speaking based on the obtained results, at the early stages of hydration (after 7 days) the compressive strength decreases. Also the compressive strength of all specimens of group 1B decreases as the MP replacement level increased. The major component of MP is calcium oxide as indicated in table 6. The CaO reacts with mixing water and forms $Ca(OH)_2$. The formed $Ca(OH)_2$ results in lowering the early strength of concrete [24].
The strength of the MP sample after 28 days of hydration increases with increase the amount of the MP up to 10% MP replacement, then decreases. The maximum increament of compressive strength of 6% as compared to the control mix has been achieved at 10% MP as replacement of cement as depected in fig 7. However the main component of MP is CaO, but it also contains a remarkable amount of SiO_2. As explained earlier the CaO reacts with mixing water to form $Ca(OH)_2$, which in turn reacts with SiO_2 to produce more CSH gel in the structure, therby enhancing the compressive strength. At high level of MP more than 10%, the generated $Ca(OH)_2$ cannot find sufficient SiO_2 to react with. As well known the excess amount of MP deteriorate the compressive strength of the concrete. Flexural Strength for different Mixes of Phase 1-B is expressed in fig. 8. As shown in fig, The Flexural Strength results have the same trend as that of the compressive strength after 28 days of hydration. The maximum gain in flexural strength was obtained at 10% MP replacement.

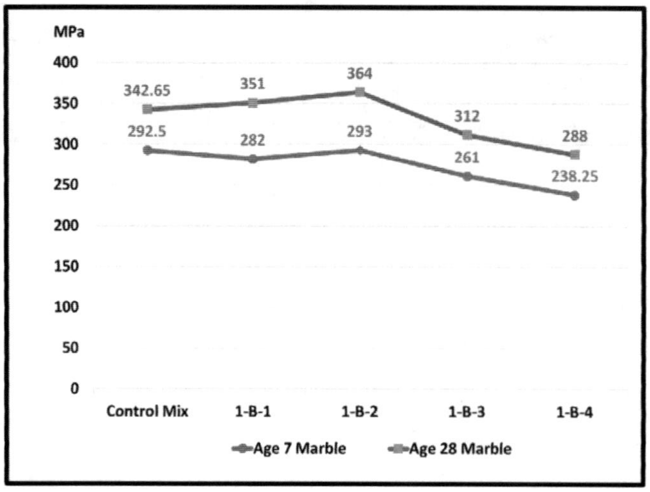

Figure 6: Compressive Strength for Different Mixes of Phase 1-B

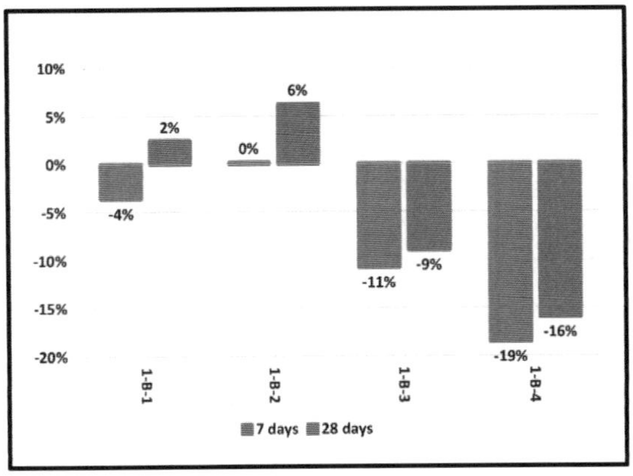

Figure 7: % Reduction/Increase of Compressive Strength for Different Mixes of Phase 1-B

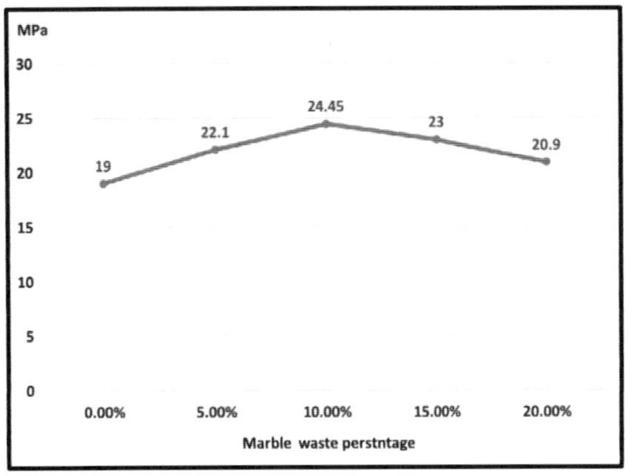

Figure 8: Flexure Strength for Different Mixes of Phase 1-B

The SEM micrograph of 1-B-4 sample is depicted in fig. 9. The SEM image exhibited a lack of hydration products espicially CSH gel responsible for the consistency and the mechanical performance of the structure. This in turn led to a deterioration in the packing and homoginity of the structure, therby deteriorate the mechanichal performance. It is apperanet that the MP particles is not tightly bound to the concrete matrix, and located inividually.

Fig. 9: SEM image of 1-B-4 sample

4.3. Phase 2

This phase of the proposed work comprises study the effect of CNTs on mechanical properties and microstructure of samples 1A-4 and 1B-4. As confirmed earlier, the addition of 20% CW and 20% MP to the concrete as replacement of cement led to deterioration in mechanical properties. So the main objective of this phase is to promote the mechanical properties using CNTs. A fixed ratio of CNTs (0.05%) was incorporated in concrete matrix.

This proportion ratio was determined based on previous studies [25]. The compressive strength values of 1A-4 and 1B-4 at early and later stages of hydration are shown in fig. 10, while fig. 11 represents Reduction/increment in Compressive strength. It was observed that at early stage of hydration the compressive strength as compared to control mix was decreased to 4.1 MPa and 2.2 MPa for 2A-4 and 2B-4 samples respectively, this may be explained as; at the early stage of curing, the hydration reaction was incomplete and still in its early stages, so the hydration products weren't strongly formed, especially C–S–H, which is responsible for the strength gain, especially the CNT has no chemical contribution in enhancing the concrete strength [26].

With further increment in a hydration time the compressive strength has been increased, this could be due to the hydration reaction was in its final stages and almost completed, which led to dense and strong hydration products in addition to the CNTs inhibiting the propagation of micro and nano cracks [27].

The flexural strength results are graphically presented in fig.12 The flexural strength results showed the same trend as that of the compressive strength at both early and later stages of hydration.

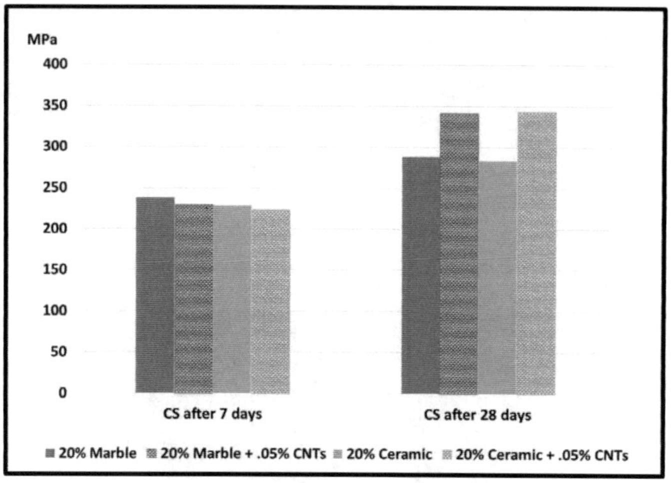

Figure 10: Compressive Strength for Different Mixes of Phase 2

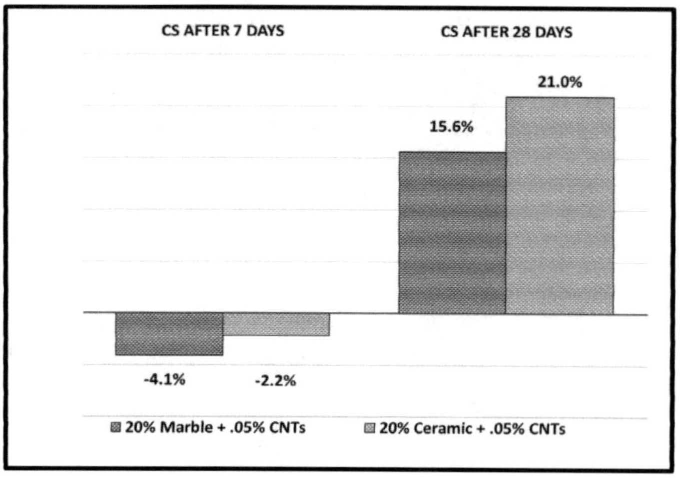

Figure 11: Reduction/increment in Compressive strength after using CNTs as an additive to (20% marble or 20% ceramic (cement replacement)

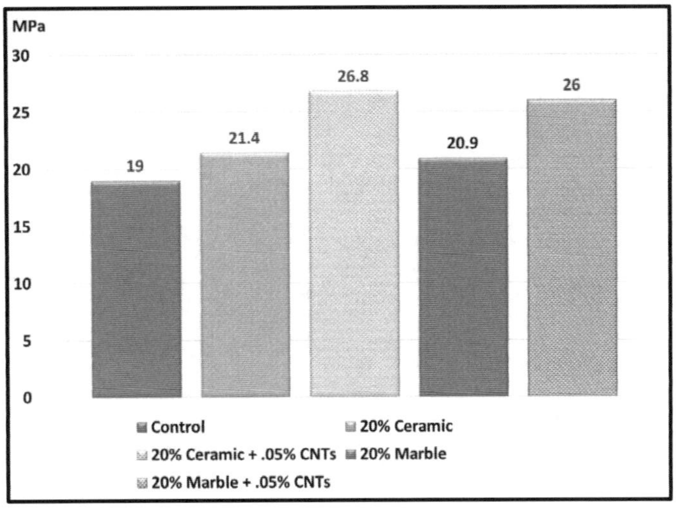

Figure 12: Flexure Strength for Different Mixes of Phase 2

The SEM micrographs of 2-A-4 sample (Fig. 55a) and 2-B-4 sample (Fig. 13) are depicted in fig. 9 The microstructure of 2-A-4 and 2-B-4 contain CSH, CNTs some CW and MP particles respectively. It was observed that the CSH gel for both samples become predominent, which led to an compact and homogenious structures with less apperant pores [28]. Also The incorporation of CNTs bridged the micro cracks. The previous reported reasons led to a conspicuous improvement in the mechanical properties.

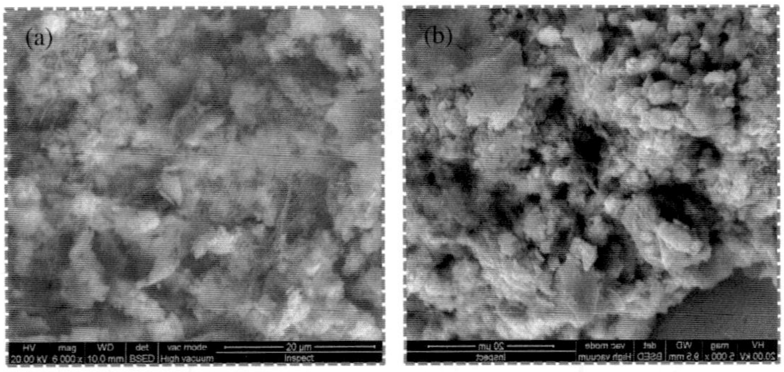

Fig. 13: SEM image of 1-B-4 sample

4.4.Phase 3

The third phase of this work is concentrated on evaluating the effect of GO on the mechanical performance and microstructure characteristics of concrete samples 3A-4 and 3B-4
The compressive strength data are presented in fig. 14, while the reduction/enhancement ratios compared to the control mix are shown in fig. 15.
From the presented figures it is obvious that the mechanical performance of concrete reinforced with 0.05 wt% of GO has the same trend as its counterpart concrete reinforced with CNTs, where after 7 days of curing, the replacement of cement by 20 wt.% of CW and MP respectively, caused a reduction in the compressive strength by 3.1 and 1.4 % for CW and MP concrete samples respectively, although the presence of GO.
Whereas after 28 days of curing, the compressive strength of CW and MP concrete samples in presence of GO increased by 11.7 and 14.6% respectively, compared to the control mix.
Which can be attributed to the lack of hydration products especially C–S–H phase that is responsible for the strength of concrete due to the incomplete hydration reaction after 7 days of curing.
While after 28 days, the hydration products were strongly formed due to the closer from the completion of the hydration reaction in addition to the presence of GO which has two functions in improving the compressive strength; one through chemical interactions between oxygenated functional groups and hydration products that led to promoting the hydration process and formation of strong interfacial forces, second is the high aspect ratio, 2D sheet-like structure of GO and its wrinkled morphology that led to significant improvement in mechanical interlocking
The flexural strength results are graphically presented in fig.16, while the reduction/enhancement ratios compared to the control mix are shown in fig. 17.
The flexural strength results exhibited the same behavior as that of the compressive strength at both 7 and 28 days of curing.
.

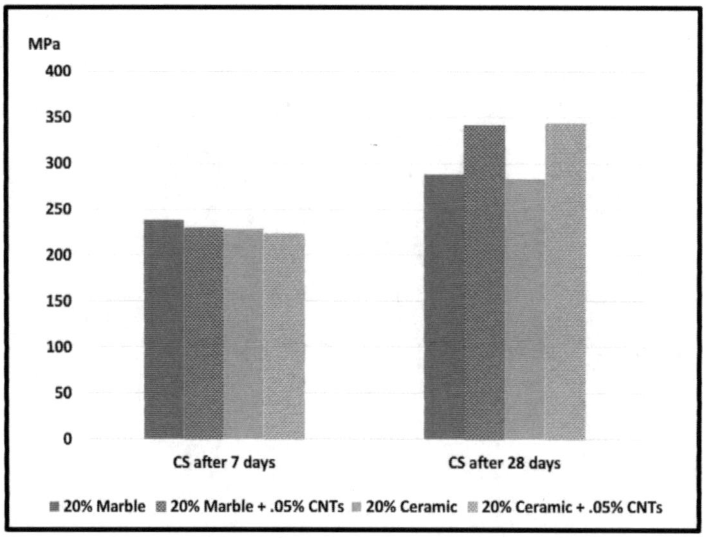

Figure 14: Compressive Strength for Different Mixes of Phase 3

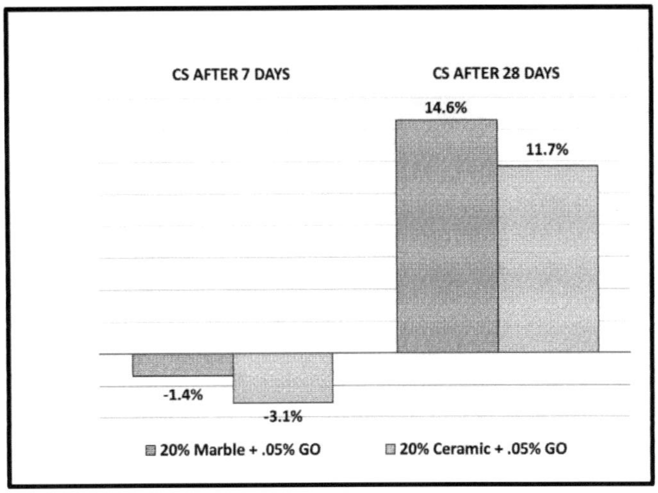

Fig. 15: Reduction/increment due to using Graphene Oxide (GO) as additive to (20% marble OR 20% ceramic (cement replacement))

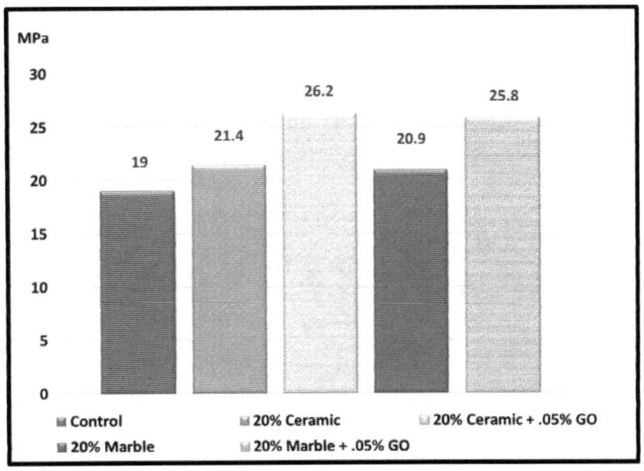

Figure 16: Flexure Strength for Different Mixes of Phase 3

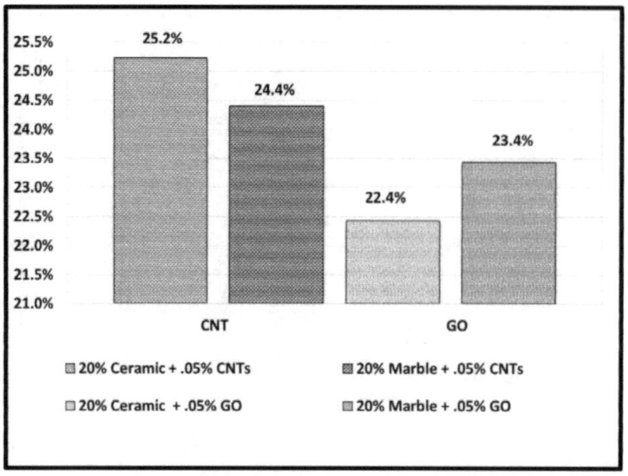

Fig. 17: Reduction/increment in Flexure strength due to using CNTs OR GO as an additive to (20% marble OR 20% ceramic (cement replacement))

The SEM micrographs of fig. 18a and 18b represent the 3-A-4 and 3-B-4 samples respectively. It is clear that the presented images contain the ordinary hydration phases in addition to GO sheets and some CW and MP particles. It was observed that the CSH gel for both samples becomes the most widespread phse in the matrix, which in turn resulted in a more dense and consistent structures with less microcracks. Also the incorporation of GO acts as a nucleation centers for the growth of hydration producs in addition to the interlocking behavior that prevent the propagation of crscks.

It also apparent that, there is no separation between the aggregates and other concrete components. This in turn confirms the developments obtained in the mechanical properties.

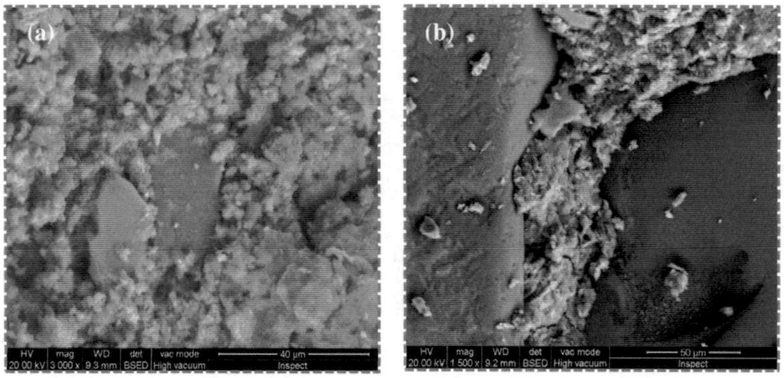

Fig. 18: SEM image of a) 3A-4 sample and b) 3B-4 sample

5. CONCLUSION:

1- MWCNTs and GO have been utilized to improve the mechanical properties of concrete containing CW and MP.
2- An enhancement up to 3% in compressinve strength has been recogonized for 10% CW relacement of cement content after 28 days of hydration.
3- The deterioration in the mechanical stength, at 20%CW is confirmed using SEM. The SEM image exhibited a loose structure with agglomerated CW particles which have irregular and angular morphology that analogous to the morphology of cement grain.
4- The strength of the MP sample after 28 days of hydration increases with increase the amount of the MP up to 10% MP replacement, then decreases.
5- The compressive strength values of 1A-4 and 1B-4 at early stages of hydration was decreased. With further increment in a hydration time the compressive strength has been increased, this could be due to the hydration reaction was in its final stages and almost completed.
6- The effect of GO on the mechanical performance and microstructure characteristics of concrete samples 3A-4 and 3B-4 has been examined. The compressive strength of CW and MP concrete samples in presence of GO increased by 11.7 and 14.6% respectively, compared to the control mix.

REFERENCES

[1] Aly M, Hashmi M S J, Olabi A G, Messeiry M, Abadir E F and Hussain A I 2012 Effect of colloidal nano-silica on the mechanical and physical behaviour of waste-glass cement mortar *Materials and Design* **33** 127–35
[2] Meyer C 2009 The greening of the concrete industry *Cement and Concrete Composites* **31** 601–5
[3] Rashad A M 2013 A preliminary study on the effect of fine aggregate replacement with metakaolin on strength and abrasion resistance of concrete *Construction and Building Materials* **44** 487–95
[4] Umapathy U, Mala C and Siva K 2014 Assessment of Concrete Strength Using Partial Replacement of Coarse Aggregate for Wast Tiles and Cement for Rice Husk Ash in Concrete **4** 72–6
[5] Dayalan J and Beulah M 2014 Effect of Waste Materials in Partial Replacement of Cement Fine Aggregate and Course Aggregate in Concrete 33–6
[6] López V, Llamas B, Juan A, Morán J M and Guerra I 2007 Eco-efficient Concretes: Impact of the Use of White Ceramic Powder on the Mechanical Properties of Concrete *Biosystems Engineering* **96** 559–64
[7] Ergün A 2011 Effects of the usage of diatomite and waste marble powder as partial replacement of cement on the mechanical properties of concrete *Construction and Building Materials* **25** 806–12
[8] Kaplan H 2007 Influence of marble and limestone dusts as additives on some mechanical properties of concrete *Scientific Research and Essays* **2** 372–9
[9] El-Haggar S M 2007 *Sustainable Industrial Design and Waste Management*
[10] Shah S P, Hou P and Konsta-Gdoutos M S 2015 Nano-modification of cementitious material: Toward a stronger and durable concrete *Journal of Sustainable Cement-Based Materials* **5** 1–22
[11] Amanullah M and Al-Tahini A M 2009 Nano-technology- its significance in smart fluid development for oil and gas field application *Society of Petroleum Engineers - SPE Saudi Arabia Section Technical Symposium 2009*
[12] Patil R and Deshpande A 2012 Use of nanomaterials in cementing applications *Society of Petroleum Engineers - SPE International Oilfield Nanotechnology Conference 2012*
[13] Ershadi V, Ebadi T, Rabani A ., Ershadi L and Soltanian H 2011 The Effect of

Nanosilica on Cement Matrix Permeability in Oil Well to Decrease the Pollution of Receptive Environment *International Journal of Environmental Science and Development*

[14] Li H, Xiao H G, Yuan J and Ou J 2004 Microstructure of cement mortar with nano-particles *Composites Part B: Engineering*

[15] Zhang R, Cheng X, Hou P and Ye Z 2015 Influences of nano-TiO2 on the properties of cement-based materials: Hydration and drying shrinkage *Construction and Building Materials*

[16] Wang Q, Wang J, Lu C X, Liu B W, Zhang K and Li C Z 2015 Influence of graphene oxide additions on the microstructure and mechanical strength of cement *Xinxing Tan Cailiao/New Carbon Materials*

[17] Siddique R and Mehta A 2014 Effect of carbon nanotubes on properties of cement mortars *Construction and Building Materials* **50** 116–29

[18] Morsy M, Yahia I S, Zahran H Y, Meng F and Ibrahim M 2019 Portable and Battery Operated Ammonia Gas Sensor Based on CNTs / rGO / ZnO Nanocomposite **48** 7328–35

[19] Morsy M, Helal M, El-Okr M and Ibrahim M 2015 Preparation and characterization of multiwall carbon nanotubes decorated with zinc oxide *Der Pharma Chemica* **7** 139–44

[20] Salama T M, Morsy M, Shahba R M A and Mohamed S H 2019 Synthesis of Graphene Oxide Interspersed in Hexagonal WO 3 Nanorods for High-Efficiency Visible-Light Driven Photocatalysis and NH 3 Gas Sensing **7** 1–14

[21] Ibrahim M 2018 Low Cost Alcoholic Breath Sensor Based on SnO 2 Modified with CNTs and **73** 1437–43

[22] Mohammadhosseini H, Lim N H A S, Tahir M M, Alyousef R, Samadi M, Alabduljabbar H and Mohamed A M 2020 Effects of Waste Ceramic as Cement and Fine Aggregate on Durability Performance of Sustainable Mortar *Arabian Journal for Science and Engineering* **45** 3623–34

[23] S. El-Dieb A, R. Taha M and I. Abu-Eishah S 2019 The Use of Ceramic Waste Powder (CWP) in Making Eco-Friendly Concretes *Ceramic Materials - Synthesis, Characterization, Applications and Recycling*

[24] Ban C C, Sern L J and Jasme N 2019 The mechanical strength and drying shrinkage behavior of high performance concrete with blended mineral admixture *Jurnal Teknologi* **81** 59–67

[25] Gao Y, Jing H, Zhou Z, Shi X, Li L and Fu G 2021 Roles of carbon nanotubes in reinforcing the interfacial transition zone and impermeability of concrete under different water-to-cement ratios *Construction and Building Materials* **272** 121664

[26] El-Feky M S, El-Khodary S A and Morsy M 2019 Optimization of hybrid cement composite with carbon nanotubes and nano silica using response surface design *Egyptian Journal of Chemistry* **62**

[27] Morsy M, Shoukry H, Mokhtar M M, Taha N A and Morsy M S 2020 Systematic investigation into mechanical strength, pore structure and microstructure of high performance concrete incorporating nano-hybrids *IOP Conference Series: Materials Science and Engineering* **956**

[28] Morsy M S, Rashad A M, Shoukry H, Mokhtar M M and El-Khodary S A 2020 Development of lime-pozzolan green binder: The influence of anhydrous gypsum and high ambient temperature curing *Journal of Building Engineering* **28** 101026

Magnetic field induced multichannel tunable filter properties of photonic band gap materials

Arafa H Aly[(1,*)], Suneet K Awasthi[2], Asmaa M Mohamed[1], Walied Sabra[1], M Mobarak[3], Z.S.Matar[4] and A S Shalaby[3]

[1]TH-PPM Group, Physics Department, Beni-Suef University, Beni Suef, Egypt.

[2]Department of Physics and Material Science and Engineering, Jaypee Institute of Information Technology, Noida 201304, UP, India.

[3]Physics Department, Beni-Suef University, Beni Suef, Egypt.

[4]Umm-Al-Qura University, Faculty of Applied Science, Department of Physics, Mecca, Saudi Arabia

(*) corresponding author: arafa.hussien@science.bsu.edu.eg

Abstract: we have explored the tunable multichannel characteristics of one-dimensional (1D) plasma photonic structure in presence of static magnetic field applied externally parallel and anti-parallel to the direction of propagation under normal incidence. We have theoretically examined the transmission characteristics of the binary photonic design consisting of alternate layers of CaF_2 and magnetic cold plasma layers by means of transfer matrix method (TMM) which is amongst the one of the popular techniques for simulating 1D multilayer periodic structures based on MATLAB. The proposed structure possess N-1 number of distinct transmission peaks, each of unit transmission called as transmission channels in transmission spectra for given the number of periods N >1. In this study N varies from 2 to 6 in steps of 1 to get 1 to 5 transmission channels respectively. Further we have also investigated how these transmission channels can be repositioned inside PBG by applying the static magnetic field (B) externally under right hand polarization (RHP) and lift hand polarization (LHP) configurations for given N at B = 0T, 0.02T, 0.04T and 0.06T. The reason behind the existence of these transmission channels is due to the superposition of evanescent and propagating waves inside plasma and dielectric layers respectively. The existence of these transmission channels inside PBG is different from the transmission bands which are formed due to the interference of forward and backward propagating waves. Besides this the tuning sense of the transmission channels at fixed B applied parallel or anti parallel to direction of periodicity, by changing the angle of incidence corresponding to TE and TM polarization case has also been examined to get some more useful and interesting tunable multichannel characteristics of the current design which cannot be obtained in conventional PPCs.

1. Introduction

The discovery of pioneering work of photonic band gap (PBG) and photon localization by two scientists Yablonovitch and John in 1987 [1, 2] attracted the attention of various research groups worldwide. PCs have amazing optical properties to control the propagation light due to the existence of PBGs which result from multiple Bragg scattering of light from the interface separating two media of the multilayer periodic structure [3-10]. PBGs are optical counterparts of the electronic band gaps (EBGs) in semiconductors [11-14]. PCs are periodic structures in which materials layers of different refractive indices like dielectric, polymers, metals and superconductor modulate periodically [15-22]. These engineered structures have variety of applications such as omnidirectional reflector, high Q-microcavties, all-optical switches, optical transistors, waveguides, filters, biosensors and others [23-32]. Beside this PCs can also be used for light trapping applications by creating deformation in the periodicity of the photonic structures [33-37]. This deformation generates a resonant tunneling mode, also called defect mode inside the photonic band gap (PBG) of structure and capable to exhibit a material-specific response profile of defect mode to external stimuli [38, 39]. The formation of the defect mode capable of sensing very minute change in the defect layer refractive index which are received considerable attention for designing of high performance photonic biosensors due to availability of easier and advanced photonic fabrication techniques to realize such structures [40-46]. Such 1D defective PC structures are responsive structures and are exploring new research field based on photonic sensing and technology due to their low cost and the easier fabrication techniques as compared to 2D and 3D defective PCs [47-49].

On the other hand PCs based on plasma as proposed by Hojo *et al.* first time in 2004 possess PBG which can be guided by altering the parameters associated with plasma like density and thickness of constituent material layers in microwave region [50]. They name this structure as plasma photonic crystals (PPCs), whose PBG characteristics can be tuned contrary to the PBGs characteristics of the conventional PCs. Later on PPC has been modified by replacing non-magnetic plasma layer with magnetic cold plasma layer to modify the structure. These modified structures with tunable PBG characteristics in presence of an externally applied magnetic field are termed as magnetically tunable plasma photonic structures (MPPCs) [51-53]. Such structures have attracted more interest of the researchers worldwide due to their externally tunable magnetic field dependent EM properties and the ability to control density of electron in plasma [54]. Moreover a tremendous property of arising one or more resonant peaks each of unit transmission dependent on period number, without breaking the periodicity of the structure has been found in MPPC due to the interaction between forward and backward decaying waves inside magnetic cold plasma layer, called as evanescent waves and propagating waves in dielectric layers [39, 55]. Awasthi *et. al.* used such properties to investigate the magnetic field induced tunable multi-channel filter properties of 1D photonic structure whose period is composed of magnetized cold plasma layer which is sandwiched between two identical layers of quartz glass in the presence of evanescent wave [56] whereas the optical properties of 1D MPPC associated with dispersion relation and PBGs corresponding to TM polarization with arbitrary magnetic declination have been theoretically investigated by Zhang *et al.* [57] Though in MPPCs without defect a tunable resonant mode of unit transmission can be found inside PBG, Kong *et al* explored the idea of doping of semiconductors to dope conventional 1D PC by magnetic cold plasma layer and suggested how a defect mode corresponding to the TE polarized wave inside PBG of 1D PPC doped with magnetic cold plasma [58] can be guided. Some other interesting optical properties of PC based on magneto-optical effects are observed by Bin *et al.* [59]. Besides this Qi *et. al.* studied how the application of periodically varying external magnetic field on a single plasma layer can also be used to get some fascinating PBG properties of great importance [60]. Recently, Awasthi *et al.* studied magnetic field induced tunable transmission and reflection characteristics of 1D hetrostructures composed of MPPCs in microwave frequency region capable of filtering 3N-3 transmission channels for period number greater than one [61]. Aforementioned research work based on MPPC whose PBG characteristics can be externally controlled have attracted great attention in past few years due to their technological utilization in the field of photonic microwave engineering.

All the above mentioned excellent piece of research works carried by various research groups' worldwide show the importance of externally tunable PBG properties of various 1D MPPC structures composed binary or ternary layers of magnetic cold plasma and dielectric materials. The purpose of this article is to extend the findings of Chang *et al.* [62] and Li *et al.* [63] by exploring the possibility of 1D binary MPPC structure to be used as tunable multichannel filter in microwave frequency region (3 – 7) GHz in presence of an external magnetic field in two different configurations. As far as our knowledge in this field no one has explored the multiple channels filtering capabilities of 1D binary MPPC structure between microwave frequency range 3 GHz to 7 GHz. In this paper, we have investigated the tunable transmission properties of microwave passing through the 1D MPPC whose period is composed alternating layers of two different materials with the help of standard transfer matrix method (TMM) theoretically [64-69].

2. Theoretical model and formulation

In this section we discuss the schematic of the proposed 1D plasma photonic crystal (PPC) [air/$(AB)^N$/air] consisting of alternating layers A and B of material CaF_2 and magnetized cold plasma respectively. The entire structure has been placed inside external magnetic field B as shown in Fig. 1.

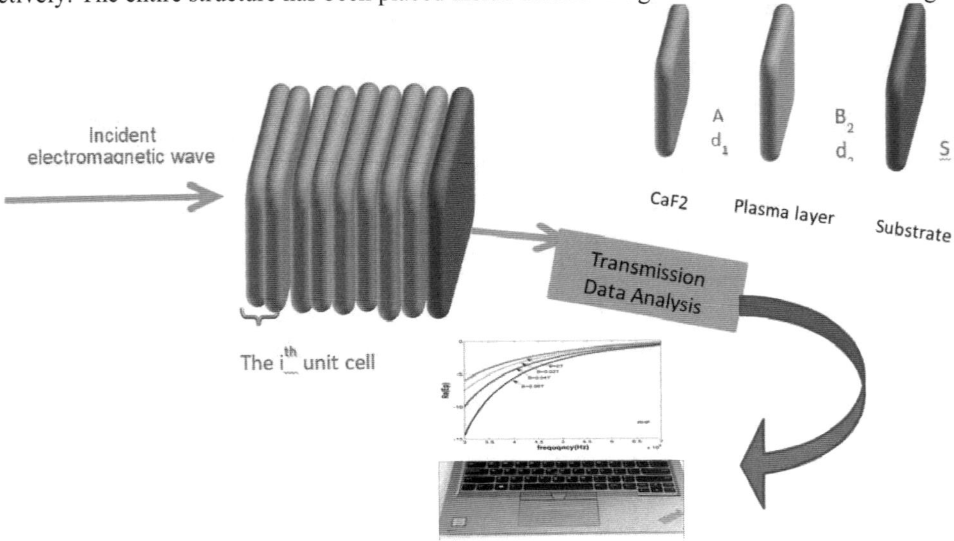

Fig.1. Schematic of the proposed design of 1 PPC $(AB)^N/S$ under the influence of B in RHP arrangement. Here A, B and S are used to represent dielectric layer of CaF_2, cold plasma layer and substrate respectively.

The transfer matrix formulation [70] (TMM) has been adopted for simulations of the results. The period number of the structure is N. Our structure is immersed in air. The thickness and refractive index of layers A and B are represented by d_1 & d_2 and n_1 & n_2 respectively. The refractive index of plasma layer B is further defined as $n_2 = \sqrt{\varepsilon_2}$, here ε_2 is the relative permittivity of plasma layer. The magnetic permeability of both the media is assumed to be 1 because of non-magnetic nature of their magnetic permeability. Suppose a plane wave is allowed to fall on 1D photonic structure at incident angle θ_o with respect direction of periodicity (which is z-axis in our case) from air of refractive index 1. The electric field (E) and magnetic field (H) vectors of s- [transverse electric (TE)] and p- [transverse magnetic (TM)] polarized incident microwave are perpendicular to xz plane respectively.
In order to connect electric and magnetic field vectors of incident microwave at every interface of the structure we use TMM as [71]

$$M_i = \begin{pmatrix} \cos\gamma_i & -\dfrac{i}{q_i}\sin\gamma_i \\ -iq_i\sin\gamma_i & \cos\gamma_i \end{pmatrix}, \qquad (1)$$

where $\gamma_i = \dfrac{\omega}{c}\sqrt{\varepsilon_i}\sqrt{\mu_i}\, d_i \cos\theta_i$, the speed of light in air is c, angular frequency of the incident microwave is ω, thickness of the ith layer is d_i, the ray angle inside i^{th} layer is θ_i. The relative permittivity and permeability of ith layer is represented by ε_i and μ_i respectively. The value of $\cos\theta_i$ in terms of incident angle θ_0 is obtained with the help of Snell's law as $cos\,\theta_i = \sqrt{1-\frac{(n_o\, sin\,\theta_o)^2}{\varepsilon_i\mu_i}}$. Here n_o is the refractive index of ambient medium which is air in our case. For s- and p- polarized waves the value of q_i will be $\frac{\sqrt{\varepsilon_i}}{\sqrt{\mu_i}}\cos\theta_i$ and $\frac{\sqrt{\mu_i}}{\sqrt{\varepsilon_i}}\cos\theta_i$ respectively.

In the proposed 1D PPC design as shown in Fig.1, TMM has been used as a most effective tool to analyze its optical properties. For connecting electric and magnetic fields of first and last layers of whole structure as depicted in Fig.1 TMM has been utilized as

$$\begin{pmatrix} E_1 \\ H_1 \end{pmatrix} = (M_A M_B)---(M_A M_B)\begin{pmatrix} E_{n+1} \\ H_{n+1} \end{pmatrix}$$
$$= M^N \begin{pmatrix} E_{n+1} \\ H_{n+1} \end{pmatrix} = \begin{pmatrix} m_{11} & m_{12} \\ m_{21} & m_{22} \end{pmatrix}\begin{pmatrix} E_{n+1} \\ H_{n+1} \end{pmatrix} \quad (2)$$

Where m_{11}, m_{12}, m_{21} and m_{22} are representing the elements of total transfer matrix of whole 1D PPC structure in presence of an external magnetic field B. The transmission coefficient of the proposed structure is defined as

$$t = \frac{2q_o}{m_{11}q_o + m_{12}q_o q_{n+1} + m_{21} + m_{22}q_{n+1}} \quad (3)$$

Here q_o and q_{n+1} are corresponding to the incident ($z < L$) and exit ($z > L$) media of the structure respectively. The total length of the proposed design is L. For s-polarized wave $q_o = q_{n+1} = \sqrt{\varepsilon_o/\mu_o}\cos(\theta_o)$ whereas for p- polarized wave $q_o = q_{n+1} = \sqrt{\mu_o/\varepsilon_o}\cos(\theta_o)$.

The transmittance of the structure can be obtained with the help of following relation as

$$T = \frac{q_{n+1}}{q_o}|t|^2. \quad (4)$$

Next the efforts are given to discuss how static magnetic field influences the behavior of permittivity function of cold plasma layer. The ω and B dependent permittivity function of cold plasma layer is defined by Eq. (5) as [39, 50, 59, 72]

$$\varepsilon_B(\omega, B) = 1 - \frac{\omega_P^2}{\omega^2\left[\left(1-\frac{i\gamma}{\omega}\right) \mp \frac{\omega_c}{\omega}\right]} \quad (5)$$

where symbols ω_p, ω_c and γ are used to represent plasma frequency, gyro-frequency and collision frequency respectively of cold plasma. The negative and positive signs prior to ω_c be used to represent the application of externally applied static magnetic field along the direction of periodicity and opposite to direction of periodicity respectively. The application of external magnetic field along the direction of periodicity is known as right-hand polarization (RHP) whereas opposite to the direction of periodicity is known as left-hand polarization (LHP) configurations. The expressions of plasma and gyro frequencies of cold plasma are given in Eqs. (6) and (7) respectively as

$$\omega_{pe} = \sqrt{\frac{n_e^2 e}{m\varepsilon_o}} \quad (6)$$

$$\omega_{ge} = \frac{eB}{m} \qquad (7)$$

where m is used to represent electronic mass, e is the electronic charge, n_e is the electron density and ε_0 is electric permittivity of free space. The application of B along positive and negative z-axis represents RHP and LHP configurations respectively.

3. Results and discussions

In the proposed research work we have chosen structural parameters as $n_1 = 1.42$ (CaF_2) [68], d_1= 15 mm, d_2 = 20 mm, n_e=8×10^{17}m^{-3}, γ = 4π×10^{7} Hz. For studying B dependent tunable multi-channel filter characteristics of the proposed photonic structure $(AB)^N$ we have varied the total periods number of periods (N) of the structure $(AB)^N$ from 2 to 6 along with incident angle.

First we examined incident microwave frequency dependent properties of electric permittivity of cold plasma layer in the presence of external B in RHP and LHP configurations both as defined in Eq. (5). For this purpose we study the ω and B dependent behavior of real part of electric permittivity of plasma layer we have chosen microwave frequency region which varies from 3 GHz to 7 GHz. The reason behind the selection of frequency range 3 GHz to 7 GHz is due to fact that in this frequency range cold plasma material layer becomes as an epsilon negative material which is one of single negative material (SNG). Figs. 2(a) and 2(b) illustrate how the real part of electric permittivity [Re(ε_B)] of cold plasma material layer changes with frequency of incident radiation under the influence of B applied externally along z axis (RHP) and opposite to z axis (LHP) arrangements respectively. The red, green, blue and black dashed lines curves are at different magnetic field values as mentioned in Fig. 2. It has been found in

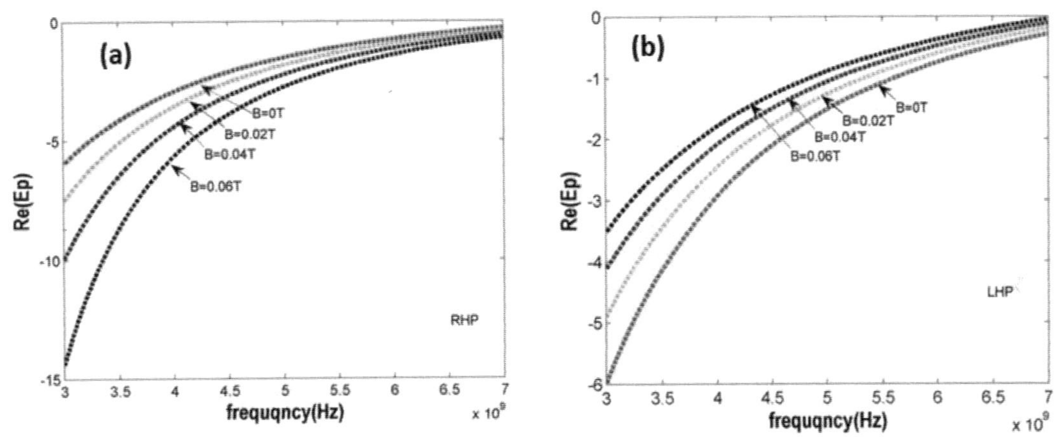

Fig. 2: Calculated ω dependent Re(ε_B) of cold plasma material layer under the influence of an external B = 0T, 0.02T, 0.04T and 0.06T shown by dashed line curve of colors red, green blue and black respectively under (a) RHP and (b) LHP arrangements.

Fig. 2 show that the increase in incident microwave frequency results the corresponding increase in the value of Re(ε_B) which reaches to its maximum value in the region of investigation. Moreover application of B from 0 to 0.06 T in steps of 0.02 under RHP arrangement results further decrease in the real part of permittivity of cold plasma layer dependent on incident microwave frequency. On the other hand electric permittivity of plasma material layer increases under the influence of B which varies from 0 to 0.06T in steps of 0.02 under LHP arrangement. In both the cases Re(ε_B) plasma layer dependent upon ω always maintain its SNG characteristics. Thus due to the different behavior of electric permittivity of plasma layer under the application of B in RHP and LHP arrangements both opens new gateway for designing of tunable SNG materials as well as associated photonic devices in microwave region. Furthermore the magnetic field B and frequency ω dependent behavior of ε_B of plasma layer is of opposite nature under RHP and LHP arrangements as shown in Fig 2. This useful property may be utilized for tuning the right and left band edges of photonic band gap and increase an additional degree of freedom for controlling of photonic band gap. Beside this negative values of ε_B of plasma material layer also indicate its metal like behavior which does not allow propagation of electromagnetic wave through it. Thus electromagnetic waves inside cold plasma layer become evanescent. In the light of above facts we can say that the ω and B dependent feature of ε_B of cold plasma layer may be very useful in designing of tunable multiple channels of different frequencies to be filtered from photonic structure consisting of magnetic field induced cold plasma layer without defect. Next the transmittance spectra of 1D PPC structure $(AB)^N$ corresponding to N=2, 3, 4, 5 and 6in absence of B at normal incidence $\theta_0 = 0^0$ has been plotted in Fig. 3 with the help of Eqs. (1) to (7). Fig. (3) shows N-1 number of distinct peaks inside PBG, called as channels in transmission spectra for given N.

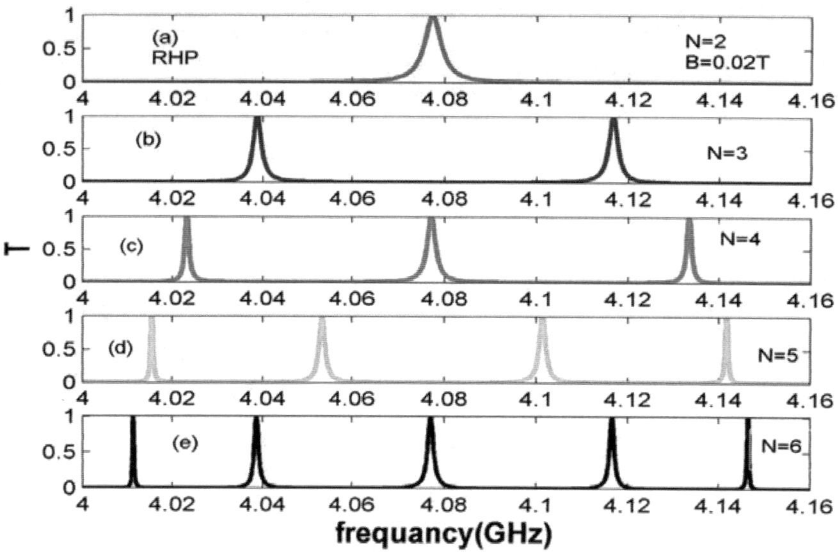

Fig. 3: Transmission spectra of 1D PPC at different value of N=2, 3, 4, 5 and 6 showing N-1 transmission channels of unit transmittance in the presence of $B = 0.02\ T$ under RHP arrangement at and $\theta_0 = 0^0$.

For N = 2, there exists only one transmission peak of unit transmission located at f = 4.076 GHz due to the superposition of evanescent waves (which arises due to forward and backward decaying waves) inside plasma layer and propagating wave inside dielectric layer. This phenomenon is not similar to the formation of transmission band resulting from superposition of forward and backward propagation waves inside dielectric layers [63]. The existence of single transmission peaks of unit transmission in N = 2 structure can also be explained on the basis of Fabry Perot cavity nature of structure $(AB)^2$ as BAB by neglecting the first layer of dielectric material A. The appearance of more than one transmission peak corresponding to structures of periods; N = 3 to N = 6 is due to the quasi Fabry-Perot cavity (FPC) nature of the structures i.e. a structure in which more than one Fabry-Perot cavities are cascaded together. In such cavities the total number of transmission peak are equal to the number of Fabry-Perot cavities cascaded together. For example $N = 6$ period 1D PPC $(AB)^6$ can be treated as quasi Fabry Perot cavity of the form $\underbrace{BA\frac{B}{2}}_{FPC-1}\underbrace{\frac{B}{2}A\frac{B}{2}}_{FPC-2}\underbrace{\frac{B}{2}A\frac{B}{2}}_{FPC-3}\underbrace{\frac{B}{2}A\frac{B}{2}}_{FPC-4}\underbrace{\frac{B}{2}AB}_{FPC-5}$. Since this quasi Fabry-Perot cavity is formed by cascading five cavities, the total number of transmission channels would be five. The similar explanation can also be applied over other structures of periods $N = 3$ to 5. Fig. 3 also indicates that for even values of N, increase in the number of periods results corresponding increase in number of transmission channels without affecting the central transmission channel located at 4.076 GHz which is common for all even values of N. On the other hand the central transmission channel located at 4.076 GHz is absent for all odd values of N. There is one more common observation that the increase in the number of periods causes the decrease in the full width at half maxima (FWHM) of each channel of unit transmission i.e. transmission peaks get narrower. Larger the value of N more will be the N-1 transmission channels which are so close to form band pass filters.

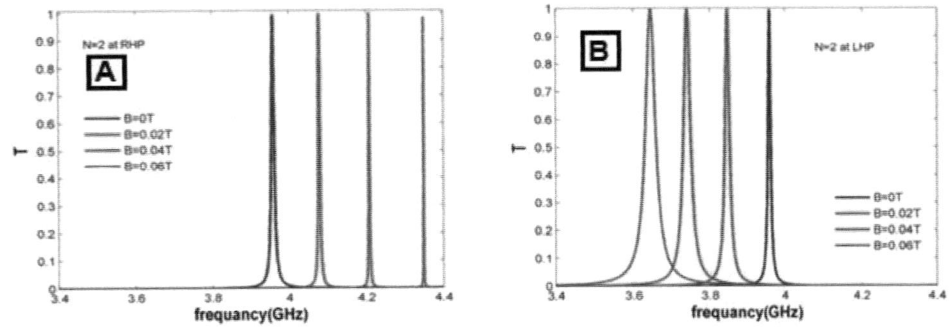

Fig. 4: Transmission spectra of 1D PPC with N = 2 and $\theta_0 = 0^0$ at different magnetic field values of 0T, 0.02T, 0.04T and 0.06T shown by solid line curve of colors blue, red, purple and green respectively under (a) RHP and (b) LHP arrangements.

Next we study the transmission properties of the proposed design with $N = 2$ periods under different values of B which varies from 0 T to 0.06 T in steps of 0.02 under RHP and LHP configuration both at normal incidence as represented in Fig. 4. It can be further noticed that under RHP arrangement, as B increases, the transmission channels shift toward higher frequency side and also transmission channels become narrower as shown in Fig. 4(a). However, mirror image behavior has been observed when the B is applied opposite to the positive direction of z axis as evident in Fig. 4(b). Thus the application of B under RHP and LHP arrangements at normal incidence can be used to change the position of transmission channels both ways as well as it is also used to control the FWHM of each transmission channels. This additional control on FWHM of each transmission channel may be utilized to control the data carrying capacity of associated channel. In addition, the separation between the transmission channels is almost same under the effect of B in RHP and LHP both configurations. Under the effect of B in LHP configuration all transmission channels have unit transmission whereas in RHP configuration transmittance is marginally reduced. In addition to all above mentioned properties the structure with period number more than two possess an additional property by which the separation between adjacent transmission channels can be either decrease with blue shift or increase with red shift under the influence of B in RHP and LHP arrangements respectively. These results are shown in Fig. 5 corresponding to $N = 3$ under the different values of B which varies from 0 T to 0.06 T in steps of 0.02 under RHP and LHP arrangements both.

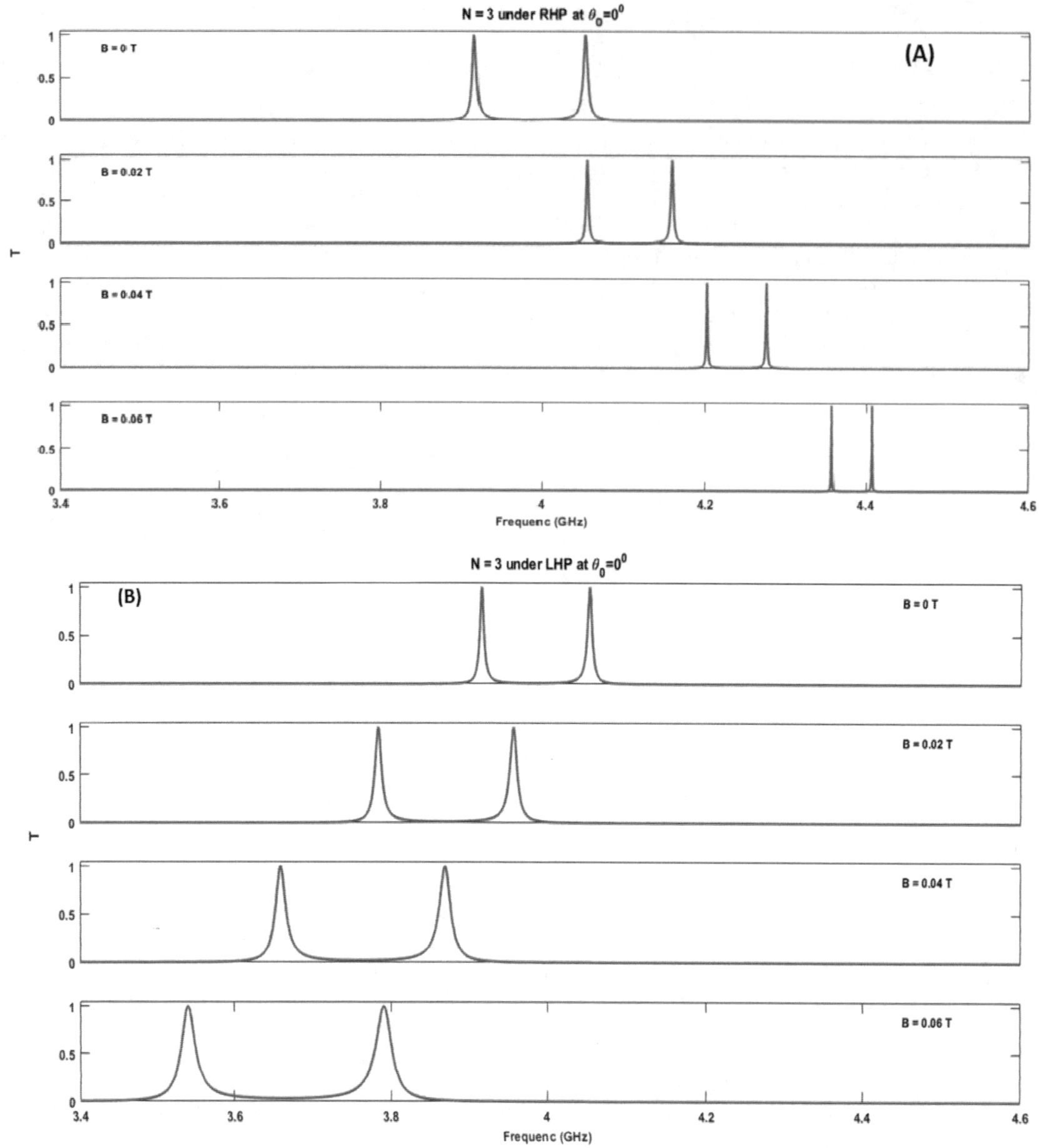

Fig. 5: Transmission spectra of 1D PPC with N = 3 and $\theta_0 = 0^0$ at different values of *B* equal to 0T, 0.02T, 0.04T and 0.06T under (a) RHP and (b) LHP arrangements.

Finally we studied how the change in incident angle θ_0 affects the central frequency of respective transmission channels. For this purpose transmission spectra of the structure corresponding to period number *N* equal to 2 at the magnetic field value of 0.0 2T under RHP configuration corresponding to four different values of incident angles of as 0°, 20°, 40° and 50° for TE and TM polarizations both have been studied as shown in Figs. 6(a) and 6(b) respectively.

IOP Conf. Series: Materials Science and Engineering 1171 (2021) 012012 doi:10.1088/1757-899X/1171/1/012012

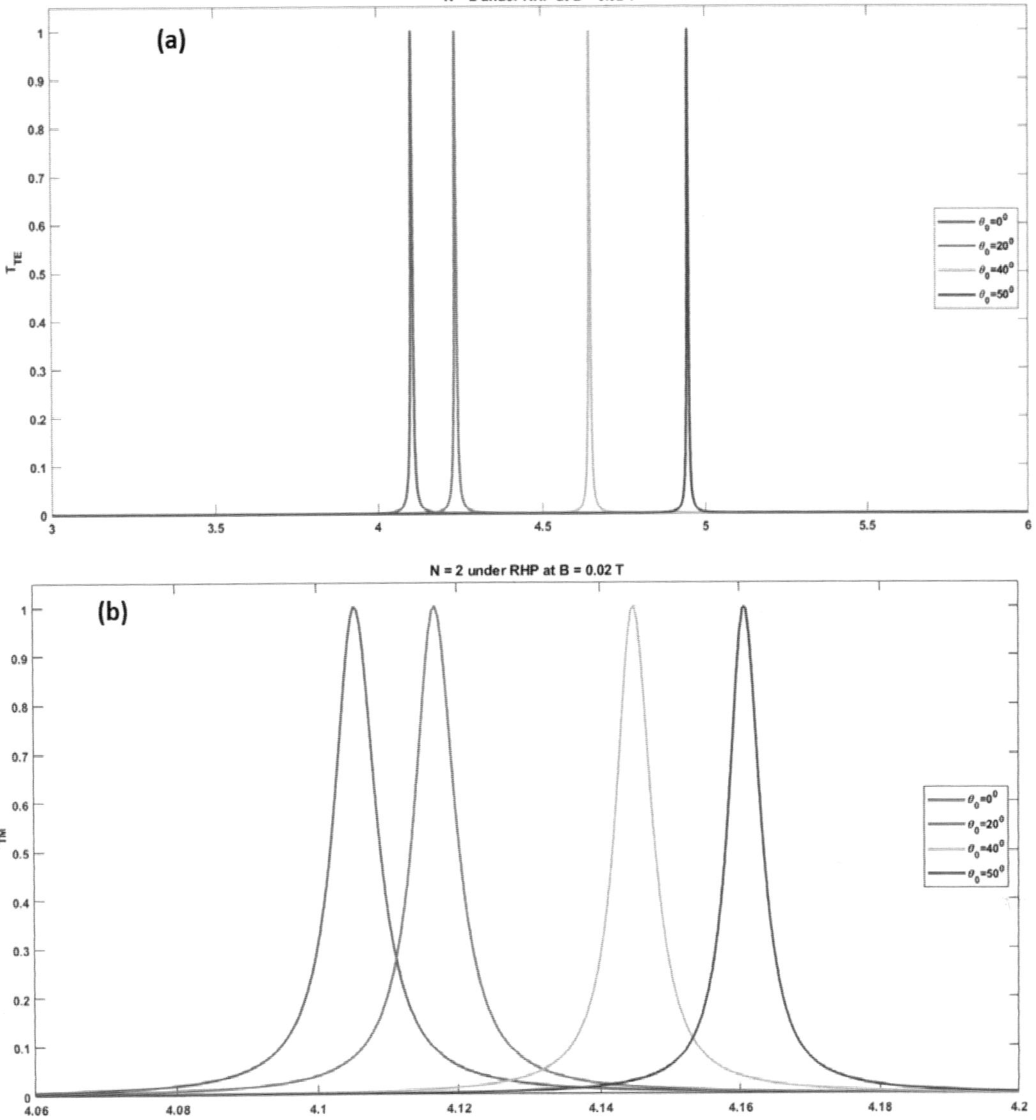

Fig. 6: Transmission spectra of 1D PPC with N = 2 and B = 0.02T under RHP arrangement at different values of incident angle $\theta_0 = 0^0$, 20^0, 40^0 and 50^0shown by solid line curves of colours; blue, red, yellow and violet corresponding to (a) TE polarization and (b) TM polarization.

It is evident from the Fig. 6 that as θ_0 increases from 0^0 to 50^0 the central frequency of each transmission channel for s- and p- polarized wave both shifted to higher frequency side. In comparison to TE and TM polarizations the shifting is more prominent in TE polarization at B equal to 0.02 T under RHP arrangement. Besides this the transmission channel corresponding to TE polarization gets narrower as θ_0 increases whereas this feature is missing in case of TM polarization. Further the transmission spectra of the structure corresponding to period number 2 at magnetic field value of 0.02 T under LHP arrangement corresponding to different values of θ_o equal to 0°, 20°, 40° and 50° for TE and TM polarizations both have also been investigated in Fig. 7. Fig. 7(a) exhibits that any increase in θ_0 for TE polarization case results the repositioning of transmission channel towards higher frequency side with decrease in the FWHM of each transmission channel at B = 0.02 T under LHP arrangement. These findings are similar to the observations found in Fig. 6(a).

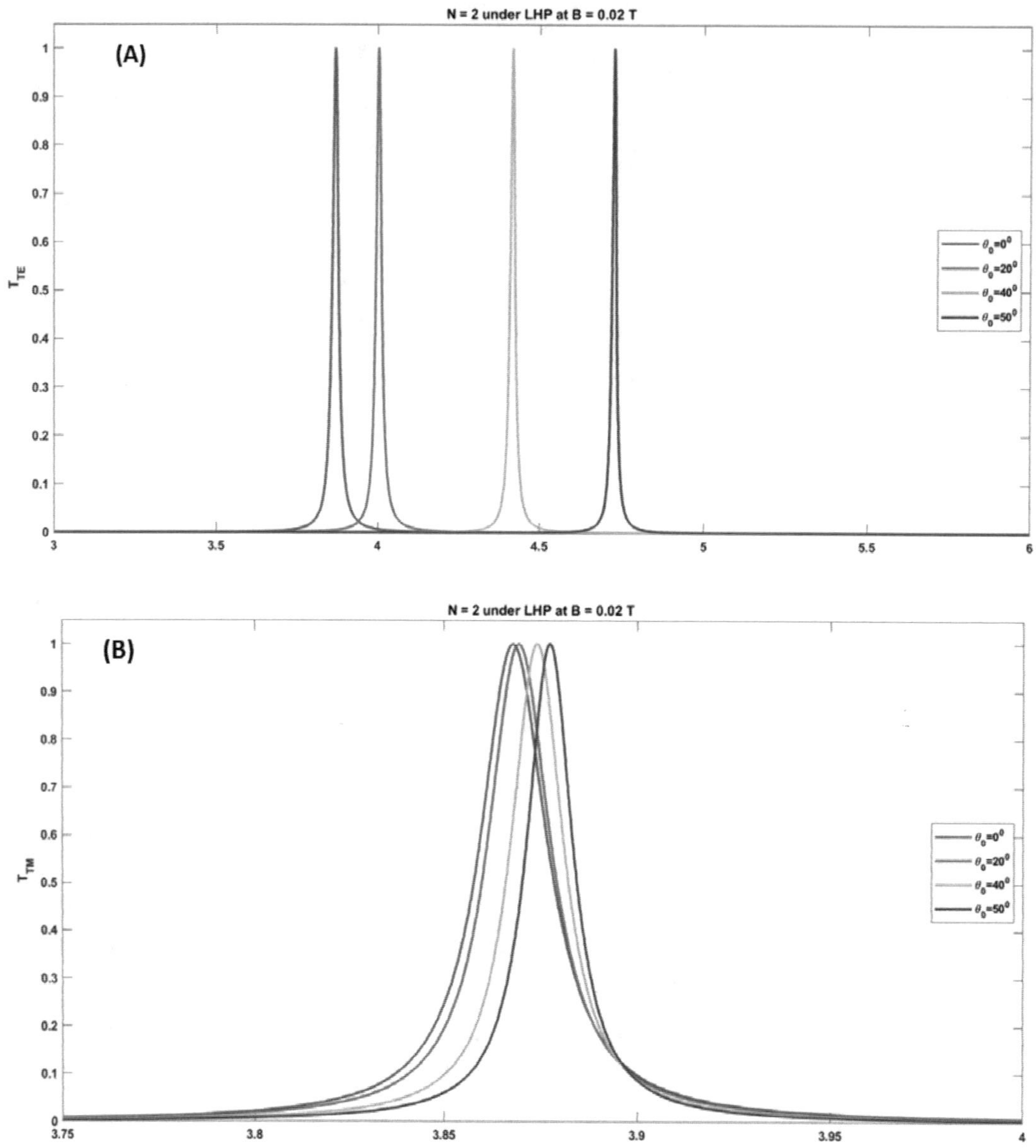

Fig.7: Transmission spectra of 1D PPC with N = 2 and B = 0.02T under LHP configuration at different values of incident angle θ_0 = 0^0, 20^0, 40^0 and 50^0shown by solid line curves of colours; blue, red, yellow and violet corresponding to (a) TE polarization and (b) TM polarization.

On the other hand the transmission spectra corresponding to TM polarization at fixed value of magnetic field of 0.02 T under LHP arrangement shows very minute impact on change in the position of transmission channel due to increase in the angle of incidence. This feature may be useful for dense division multiplexing in microwave region. Thus the application of B applied externally on 1D PPC consisting of cold plasma layer may provide an additional degree of freedom to design tunable multichannel filters of distinguished features which cannot be obtained by the means of conventional tuning approach like tuning by changing incident angle in microwave regions.

Conclusion

Here we conclude our theoretical findings of multiple transmission channels which can be guided either side of its position by means of an external magnetic field under RHP and LHP arrangements. The standard transfer matrix method has been explored to carry out numerical simulations based on MATLAB. The proposed research work can be employed to reposition the multiple transmission channels swiftly at desired frequency inside PBG externally without disturbing the experimental arrangements of the design. The external presence of B under RHP and LHP arrangements has a significant impact on ε_B of cold plasma material layer such that transmission channel frequency can be shifted to higher and lower frequency sides both. Thus we can conclude that the ω and B dependent feature of ε_B of cold plasma material layer can be utilized to design 1D PPC based tunable multichannel filter consisting of magnetized cold plasma layer without defect which may be very useful in microwave engineering applications. Apart from the application of B under RHP and LHP arrangements the transmission channel can also be guided either side of its position by changing the incident angle corresponding to TE and TM polarization separately. We have also investigated how variation in number of periods N alters the N-1 number of distinct transmission channels of unit transmission in microwave frequency region. For large values of N there are more numbers of N-1 transmission channels which are so close to form band pass filters. This study also provides an insight how to control FWHM of each transmission channel to control the data carrying capacity of associated channel which may strengthen the data load on network.

References

[1] Yablonovitch E 1987 Inhibited spontaneous emission in solid-state physics and electronics *Phys. Rev. Lett.* **58** 2059

[2] John S 1987 Strong localization of photons in certain disordered dielectric superlattices *Phys. Rev. Lett.* **58** 2486

[3] Abd El-Ghany S E, Noum W M, Matar Z, Zaky Z A and Aly A H, 2020 Optimized bio-photonic sensor using 1D-photonic crystals as a blood hemoglobin sensor *Phys. Scr.* **96** 035501

[4] Aly A H, Mohamed D, Zaky Z A, Matar Z S, Abd El-Gawaad N S, Shalaby A S, Tayeboun F and Mohaseb M 2021 Novel Biosensor Detection of Tuberculosis Based on Photonic Band Gap Materials Mat. Res. 24 (3), e20200483 https://doi.org/10.1590/1980-5373-MR-2020-0483

[5] Awasthi S 2021 Multichannel tunable polarizing filter properties of one-dimensional ternary photonic crystal containing single-negative materials *Indian J. Phys.* 1-11

[6] Nagaty A, Mehaney A and Aly A H, 2018 Influence of temperature on the properties of one-dimensional piezoelectric phononic crystals *Chinese Physics B* **27** 094301

[7] Aly A H, Mohamed D, Abd El-Gawaad N S, Matar Z S, Trabelsi Y,Mohaseb M 2020 Fano Resonance Properties in (K3C6o / HgBa2Ca2Cu3O10) 1D photonic crystal" IOP Conf. Ser.: Mater. Sci. Eng. 956, 012004

[8] Bikbaev R G, Vetrov S Y and Timofeev I V 2017 Optical Tamm states at the interface between a photonic crystal and a gyroid layer *J. Opt. Soc. Am. B-Opt. Phys.* **34** 2198-202

[9] Aly A H, and Zaky Z A 2019 Ultra-sensitive photonic crystal cancer cells sensor with a high-quality factor *Cryogenics* **104** 102991

[10] Zaky Z A, Ahmed A M and Aly A H 2021 Remote Temperature Sensor Based on Tamm Resonance *Silicon,* https://doi.org/10.1007/s12633-021-01064-w

[11] Aghajamali A 2016 Transmittance properties in a magnetized cold plasma–superconductor periodic multilayer *Appl. Optics* **55** 6336-40

[12] Aly A H, Zaky Z A, Shalaby A S, Ahmed A M and Vigneswaran D 2020 Theoretical study of hybrid multifunctional one-dimensional photonic crystal as a flexible blood sugar sensor *Phys. Scr.* **95** 035510

[13] Shiveshwari L and Awasthi S 2020 Band gap characteristics of 1D static and moving photonic crystal *Physica B* **597** 412360

[14] Zaky Z A, Ahmed A M, Shalaby A S and Aly A H, 2020 Refractive index gas sensor based on the Tamm state in a one-dimensional photonic crystal: Theoretical optimisation *Sci. Rep.* **10** 9736

[15] Aghajamali A, Zare A and Wu C-J 2015 Analysis of defect mode in a one-dimensional symmetric double-negative photonic crystal containing magnetized cold plasma defect *Appl. Optics* **54** 8602-6

[16] Aly A H and Sabra W 2013 Effective surface impedance of a high temperature superconductor Physica C: Superconductivity 495 126-129

[17] Armstrong E and O'Dwyer C 2015 Artificial opal photonic crystals and inverse opal structures–fundamentals and applications from optics to energy storage *J. Mater. Chem. C* **3** 6109-43

[18] Zaky Z A and Aly A H 2020 Theoretical Study of a Tunable Low-Temperature Photonic Crystal Sensor Using Dielectric-Superconductor Nanocomposite Layers *J. Supercond. Nov. Magn* **33** 2983–90

[19] Auguié B, Fuertes M C, Angelomé P C, Abdala N L, Soler Illia G J and Fainstein A 2014 Tamm plasmon resonance in mesoporous multilayers: toward a sensing application *Acs Photonics* **1** 775-80

[20] Aly A H, Ameen A A and Vigneswaran D 2019 Superconductor nanometallic photonic crystals as a novel smart window for low-temperature applications *J. Supercond. Nov. Magn* **32** 191-7

[21] Azzini S, Lheureux G, Symonds C, Benoit J-M, Senellart P, Lemaitre A, Greffet J-J, Blanchard C, Sauvan C and Bellessa J 2016 Generation and spatial control of hybrid Tamm plasmon/surface plasmon modes *ACS photonics* **3** 1776-81

[22] Zaky Z A and Aly A H 2021 Modeling of a biosensor using Tamm resonance excited by graphene *Appl. Optics* **60** 1411-9

[23] Awasthi S K, Malaviya U and Ojha S P 2006 Enhancement of omnidirectional total-reflection wavelength range by using one-dimensional ternary photonic bandgap material *J. Opt. Soc. Am. B-Opt. Phys.* **23** 2566-71

[24] Akahane Y, Asano T, Song B-S and Noda S 2003 High-Q photonic nanocavity in a two-dimensional photonic crystal *Nature* **425** 944-7

[25] Zaky, Z A, Aly, A H 2021, Highly Sensitive Salinity and Temperature Sensor Using Tamm Resonance, Plasmonics, https://doi.org/10.1007/s11468-021-01487-6

[26] Yanik M F, Fan S and Soljačić M 2003 High-contrast all-optical bistable switching in photonic crystal microcavities *Appl. Phys. Lett.* **83** 2739-41

[27] Ameen A A, ElSayed H and Aly A H, 2021 Towards a highly efficient air purifier using annular photonic crystals in UV regimes RSC Adv. 11 14915-21

[28] John S and Florescu M 2001 Photonic bandgap materials: towards an all-optical micro-transistor *Journal of Optics A: Pure and Applied Optics* **3** S103

[29] Zaky Z A, Sharma A, Alamri S and Aly A H 2021 Theoretical evaluation of the refractive index sensing capability using the coupling of Tamm–Fano resonance in one-dimensional photonic crystals Appl. Nanosci. https://doi.org/10.1007/s13204-021-01965-7

[30] Weily A R, Esselle K P and Sanders B C 2003 Photonic crystal horn and array antennas *Phys. Rev. E* **68** 016609

[31] Upadhyay M, Awasthi S K, Shiveshwari L, Srivastava P K and Ojha S P 2015 Thermally tunable photonic filter for WDM networks using 1D superconductor dielectric photonic crystals *J. Supercond. Nov. Magn* **28** 2275-80

[32] Aly A H, 2008 Metallic and superconducting photonic crystal *J. Supercond. Nov. Magn* **21** 421-425

[33] Aly A H, and Elsayed H A 2012 Defect mode properties in a one-dimensional photonic crystal *Physica B* **407** 120-5

[34] Bouaziz L, Nachi K, Gamra D, Lejeune M, Zeinert A, Zellama K and Bouchriha H 2015 Experimental evidence of the photonic bandgap and defect effect in Si-SiO_2 multilayer structures *Cryst. Res. Technol.* **50** 77-84

[35] Aly A H, Sayed F A and Elsayed H A 2020 Defect mode tunability based on the electro-optical characteristics of the one-dimensional graphene photonic crystals *Appl. Optics* **59** 4796-805

[36] Qutb S R, Aly A H and Sabra W 2021 Salinity and temperature detection for seawater based on a 1D-defective photonic crystal material" International Journal of Modern Physics B, Vol. 35, No. 1 (2021) 2150012 (13 pages).

[37] Sabra W, Aly A H 2021 A Comparative study of the effective surface impedance of an HTc superconducting thin film from visible to mid-IR region Opt Quant Electron **53** 416

[38] Upadhyay M, Awasthi S, Shiveshwari L, Shukla S and Ojha S 2016 Temperature-dependent tuning of photonic band gaps for wavelength-selective switching applications *Indian J. Phys.* **90** 353-8

[39] Aly A H, Elsayed H A, Ameen A A and Mohamed S 2017 Tunable properties of one-dimensional photonic crystals that incorporate a defect layer of a magnetized plasma *Int. J. Mod. Phys. B* **31** 1750239

[40] Wu C-J and Wang Z-H 2010 Properties of defect modes in one-dimensional photonic crystals *Progress In Electromagnetics Research* **103** 169-84

[41] Painter O and Srinivasan K 2002 Polarization properties of dipolelike defect modes in photonic crystal nanocavities *Opt. Lett.* **27** 339-41

[42] Aly A H, Mohamed D, AMohaseb M A, El-Gawaad N S and Trabelsi Y 2020 Biophotonic sensor for the detection of creatinine concentration in blood serum based on 1D photonic crystal *RSC Adv.* **10** 31765-72

[43] Fedele F, Yang J and Chen Z 2005 Properties of Defect Modes in One-Dimensional Optically Induced Photonic Lattices *Studies in Applied Mathematics: Special issue: Nonlinear optics* **115** 279-301

[44] Aly A H, and Elsayed H A 2017 Tunability of defective one-dimensional photonic crystals based on Faraday effect *Journal of Modern Optics* **64** 871-7

[45] Xu K-y, Zheng X and She W-l 2004 Properties of defect modes in one-dimensional photonic crystals containing a defect layer with a negative refractive index *Appl. Phys. Lett.* **85** 6089-91

[46] Aly A H and Mohamed D 2015 BSCCO/SrTiO3 One Dimensional Superconducting Photonic Crystal for Many Applications" J Supercond . Nov Magn, 18, 1699-1703

[47] Aly A H, Shaban S M and Mehaney A 2021 High-performance phoxonic cavity designs for enhanced acousto-optical interaction *Appl. Optics* **60** 3224-31

[48] Nair R V and Vijaya R 2010 Photonic crystal sensors: An overview *Progress in Quantum Electronics* **34** 89-134

[49] Sayed H and Aly A H, 2021 Salinity optical sensor by using two-dimensional photonic crystals: computational study *Mat. Sci. Eng. B.* **269** 115169

[50] Hojo H and Mase A 2004 Dispersion relation of electromagnetic waves in one-dimensional plasma photonic crystals *Journal of Plasma and Fusion Research* **80** 89-90

[51] Qi L, Yang Z, Lan F, Gao X and Shi Z 2010 Properties of obliquely incident electromagnetic wave in one-dimensional magnetized plasma photonic crystals *Physics of Plasmas* **17** 042501

[52] Kee C-S, Li S-Z, Kim K and Lim H 2003 Tunable resonant transmission of electromagnetic waves through a magnetized plasma *Phys. Rev. E* **67** 036612

[53] Gandhi A, Awasthi A K and Aly A H 2021 Biophotonic sensor design using 1D defective annular photonic crystal for the detection of creatinine concentration in blood serum RSC Adv.11, 26655

[54] Lin W-H, Wu C-J, Yang T-J and Chang S-J 2010 Terahertz multichanneled filter in a superconducting photonic crystal *Opt. Express* **18** 27155-66

[55] Shiveshwari L and Awasthi S 2015 Transmission properties of one-dimensional ternary plasma photonic crystals *Physics of Plasmas* **22** 092129

[56] Awasthi S K, Panda R and Shiveshwari L 2017 Multichannel tunable filter properties of 1D magnetized ternary plasma photonic crystal in the presence of evanescent wave *Physics of Plasmas* **24** 072111

[57] Zhang H-F, Liu S-B and Kong X-K 2012 Photonic band gaps in one-dimensional magnetized plasma photonic crystals with arbitrary magnetic declination *Physics of Plasmas* **19** 122103

[58] Kong X-k, Liu S-b, Zhang H-f and Li C-z 2010 A novel tunable filter featuring defect mode of the TE wave from one-dimensional photonic crystals doped by magnetized plasma *Physics of Plasmas* **17** 103506

[59] Mohamed A G, ElSayed A Hussein, Mehaney A, Aly A H & Walied Sabra A 2021 The transmissivity of one-dimensional photonic crystals comprising three phases nanocomposite layer for optical switching purposes Phys. Scr. 96 115504.

[60] Qi L and Zhang X 2011 Band gap characteristics of plasma with periodically varying external magnetic field *Solid state communications* **151** 1838-41

[61] Awasthi S K, Panda R, Verma A, Chauhan P K and Shiveshwari L 2019 Microwave multichannel tunable filter based on transmission and reflection properties of 1D magnetized plasma photonic crystal heterostructures *Indian J. Phys.* 1-14

[62] Chang T-W, Chien J-R C and Wu C-J 2016 Magnetic-field tunable multichannel filter in a plasma photonic crystal at microwave frequencies *Appl. Optics* **55** 943-6

[63] Li C-z, Liu S-b, Kong X-k, Zhang H-f, Bian B-r and Zhang X-y 2011 A novel comb-like plasma photonic crystal filter in the presence of evanescent wave *IEEE Trans. Plasma Sci.* **39** 1969-73

[64] Malek C, Aly A H, Alamri A, Sabra W 2021 Tunable PBGs with a cutoff frequency feature in Fibonacci quasi-periodic designs containing a superconductor material at THz region Phys. Scr. 96(10) 105501

[65] Born M and Wolf E 1980 Basic properties of the electromagnetic field *Principles of optics* **44**

[66] Aly A H, Ghany S-S A, Kamal B and Vigneswaran D 2020 Theoretical studies of hybrid multifunctional YaBa2Cu3O7 photonic crystals within visible and infra-red regions *Ceram. Int.* **46** 365-9

[67] Aly A H and Sayed H 2018 Photonic band gap materials and monolayer solar cell *Surf. Rev. Lett.* **25** 1850103

[68] Jalali T and Hessamodini M 2015 The effect of 1D magneto-photonic crystal defect mode on Faraday rotation *Optik* **126** 3954-8

[69] Awad M and Aly A H, 2019 Experimental and theoretical studies of hybrid multifunctional TiO2/TiN/TiO2 *Ceram. Int.* **45** 19036-43

[70] Aly A H, and Sayed F A 2020 THz cutoff frequency and multifunction $Ti_2Ba_2Ca_2Cu_3O_{10}$/GaAs photonic bandgap materials *Int. J. Mod. Phys. B* **34** 2050091

[71] Aly A H, Barakat S and Amin A 2020 Tunable filter based on the 1D photonic crystal within ultraviolet radiations. In: *IOP Conference Series: Materials Science and Engineering*: IOP Publishing, p 012010

[72] Guo S, Hu C and Zhang H 2020 Analysis of the features of a multifunctional device based on the regulation of the magnetic field in one-dimensional photonic crystals containing only plasma with a novel quasi-periodic structure *J. Opt. Soc. Am. B-Opt. Phys.* **37** 1996-2005

AUTHOR INDEX